Mssrs. Hartmann & Braun's
Exhibit at the Worlds Fair, Chicago 1893.*)

Mssrs. Hartmann & Braun have a splendid exhibit at Chicago. On a floor space of 100 m² in the vast Electricity Hall they have erected a magnificent pavilion wherein to house their exhibits. The design of the pavilion is Italian Renaissance enriched with ornamental plaster work. In the centre of the façade stands the statue of a female typifying Science as applied to Electricity, and tendering in her hand a laurel wreathed medallion of William Weber, the originator of our electro-magnetic standards. The interior of the pavilion is decorated with a similar magnificence. The columns of the four large entrances, which are hung with splendid draperies, and the pillars carrying the roof, form a series of bays round the pavilion walls on which the various classes of test instruments are mounted against a pale green merino background. The contrast between the elaborate and handsome polished wood work and the rich draperies, with the highly finished surfaces of the various instruments, renders the entire exhibit exceptionally magnificent. Each bay is surmounted either with a medallion of Ohm, Ampère, Volta, Watt, Coulomb, Faraday, in carton-pierre or with a shield bearing the names of Gauss, Henry, Franklin, Morse, Reis, Kohlrausch. The bust of the celebrated German Scientist W. von Siemens, a work by Hildebrand, is placed under a canopy whose draperies form a dark room for Photometry if required. Messrs. Hartmann & Braun's pavilion is, in fact, a Memorial Hall for the scientists whose researches are embodied in the instruments exhibited, as well as a site for the display of the firm's manufactures.

The pavilion includes, in addition to a complete collection of their various instruments, an electrical laboratory completely equipped for all classes of measurements (Figs. 2 and 3). The collection, of which, in most instances duplicates are shown connected and ready for use, is partly housed in beautiful plate glass cabinets and partly in table cases, and comprises: Mirror Galvanometers and suitable Telescopes for use therewith; a complete set of the Kohlrausch Instruments; Standards made in accordance with those of the Imperial Physico-Technical Laboratory; Accurate Rheostats; Bridges and Resistance-measuring apparatus for technical purposes; accurate instruments for measuring Current and determining Potential and various apparatus for Lecture Room and Demonstration purposes.

Various classes of Volt- and Ampèremeters for use on direct or alternating currents; Ohm- and Wattmeters; Accoustic Voltmeters and Supply meters; Milli-ampèremeters for medical purposes; electrical Pyrometers and Telethermometers are also shown in 14 different groups on the walls.

The Laboratory is arranged on an elevated floor in the centre of the pavilion; on white marble cased stone work and on oak tables and stands are mounted four sets of apparatus arranged for accurate determination of resistance, current and potential, insulation and capacity of cables; and in addition a dark room is provided for including a Photometer and all requisite accessories for measurements of light.

Finally, outside the pavilion is a combined testing cart and tent, and a number of well designed portable sets of apparatus for localising faults on leads and testing their insulation resistance.

The necessary current for the various apparatus, and for the general lighting throughout the pavilion, which is lit by both arc and incandescent

*) From The Electrotechnische Zeitschrift. Berlin 1893 No. 42.

lamps is supplied from a set of Pollak accumulators, which are charged from the Siemens & Halske five-wire network of mains. The lighting current is kept constant by an automatic cell switch, also made by the exhibitors.

Fig. 1.

A well got up catalogue in English, the cover ornamented with a drawing of the statue we have just described, serves as a concise descriptive guide to the various exhibits which are arranged in consecutive order round the pavilion.

Fig. 2.

In all some 330 valuable instruments in 108 different types are shown; these unquestionably form one of the most interesting collective exhibits of apparatus designed for work in any special branch of Science.

Fig. 3.

Our Exhibits at Chicago received eight awards.

NOTICE.

1. We can only consider **claims** made within fourteen days after receipt of goods.
2. All goods are **packed with the greatest care**, shipping orders are packed in tin lined cases, packing is charged at cost price and is **not** returnable.
3. All goods are sent at **consignee's own risk.**
4. **Orders from unknown correspondents**, or from correspondents who do not furnish references will only be accepted when accompanied by remittance for one third of the amount due; the remaining two thirds to be paid upon despatch of the goods.
5. As alterations and improvements are constantly being made in our apparatus, the **illustrations** and prices are liable to alteration without notice.
6. We use our best endeavours to carry out **repairs** to any of our apparatus with as little delay as possible, chargeing for same at a moderate rate, if we are unable to quote a fixed price for same beforehand, which is not always possible. Only under exceptional circumstances we can undertake the repair of apparatus not made by us.
7. Where **alterations** from our **standard patterns** are required, or **special instruments** are made to our customers specification, for which it is impossible to quote beforehand, our charges for same must be regarded as **indisputable**, and our customers rely upon our basing them as low as possible. We may add that we do **not like** to make **single** instruments to specification and can but **seldom undertake** such work.
8. The **time stated for delivery** is based upon our usual factory routine and is seldom exceeded, but is not guaranteed; we entirely decline responsibility for losses incurred through non-delivery at time stated; whether arising from errors on our part, accidents, or strikes.
9. We give the following **guarantee** with all instruments of our manufacture, namely, to repair or replace **free of charge** with new perfect instruments, all faults or errors arising within one year from date of delivery and due to faulty material or workmanship. We recognise no other compensation of any sort whatever and are not responsible for damage or loss in business arising out of these defects. We are, of course, not responsible for accidents, or damage due to fair wear and tear, unskilled manipulation or from using the instruments above their stated limits.
10. **Payment is strictly nett within thirty days**, or **draft at one month** (except in the case of Public Institutions). Small amounts will be collected by the Post Office on delivery.

We especially request our customers when ordering to give the **catalogue number of each instrument required**; unless this is done we cannot be responsible if the instruments do not comply with their requirements.

Postal Address:

"HARTMANN & BRAUN", BOCKENHEIM-FRANKFORT o. M.

Telegraphic Address:

HARTMANNBRAUN FRANKFORTMAIN.
(ONE word)

A B C code used.

References as to the accuracy and good workmanship of our apparatus are kindly permitted by the following well known scientists.

Dr. **Carl Barus**, U. S. Geological Survey . . . . **Washington.**
Dr. **E. du Bois-Reymond** . . . . **Berlin.**
Dr. **Börgen**, Imperial Observatory . . . . **Wilhelmshaven.**
Dr. **Ferd. Braun** . . . . **Tübingen.**
Dr. **Dietrich** . . . . **Stuttgart.**
Prof. **Galileo Ferraris** . . . . **Turin.**
G. **Carey Foster**, Esq., B. A., F. R. S. University College . . . **London.**
Prof. **Frithiof Holmgren** . . . . **Upsala.**
Dr. **Kittler** . . . . **Darmstadt.**
Dr. **Herm. I. Klein**, Astronomer . . . . **Cologne.**
Dr. **F. Kohlrausch** . . . . **Strassburg.**
Dr. **W. Kohlrausch** . . . . **Hannover.**
Dr. **Neumayer**, Imperial Observatory . . . . **Hamburg.**
Dr. **Quincke** . . . . **Heidelberg.**
Dr. **P. Siloff** . . . . **Warsaw.**
Dr. **W. H. Stone**, M. A.; F. R. C. P., St. Thomas Hospital . . . **London.**
Dr. **Strouhal** . . . . **Prague.**
Prof. **N. Umow** . . . . **Odessa.**
Dr. **A. von Waltenhofen** . . . . **Vienna.**
Dr. **H. F. Weber** . . . . **Zürich,**

and many others in all parts of the world.

INDEX.

———❈———

The firm is in posession of the following German Patents:

No. 36554, No. 36911, No. 39561, No. 39869, No. 51563, No. 56633, No. 56696, No. 63219, No. 66678, No. 68918, No. 69561, No. 71484, No. 74338, No. 75065, No. 75503, No. 76933, No. 77576, No. 78310,

(in addition to several applications already filed)

and a number of foreign patents.

Hartmann & Braun

Electrical Measuring Instruments

1894

PREFACE.

This edition of our catalogue is far larger than our previous issues. Our endeavours to keep pace with all the improvements in Science and Practice, and our wish to facilitate the selection of the most suitable instrument for whatever purpose required, by furnishing a slight description of the necessary manipulation, have led to this increase; which we hope will cause this catalogue to serve as a book of reference when using the instruments described.

With the exception of the copies of Standards of the Imperial Physico-Technical Laboratory, nearly all our instruments have been **entirely designed** by our firm and their associates. The usual standard types of apparatus have been remodelled and improved, to bring them up to the present day requirements for measuring instruments. Some of Prof. Kohlrausch's valuable designs have also been remodelled in accordance with his suggestions; we have always made it our special endeavour in improving and remodelling our instruments, to make them as simple and convenient as possible, so that they may be used by comparatively unskilled hands without loss of time; and for this reason we have originated several **direct reading instruments,** to which we would especially draw the attention of all practical engineers.

Especial care is given to the **calibration, adjustment** and **exact determination of Constants,** for which purpose we have a very complete and carefully designed plant, comprising 4 Dynamos for direct currents up to 1500 ampères, three sets of large accumulators and one set of several hundred small ones also capable of discharging at the rate of 2000 ampères and at potentials up to 800 volts, in addition to a direct current motor transformer for potentials up to 2000 volts and electrostatic arrangements for higher potentials. For alternating current work within very wide limits, an alternating current dynamo with 3 transformers having a range up to 1000 amps and 10000 volts is employed.

Whilst using our very best endeavours to meet all the requirements of accuracy, convenience, and mechanical perfection in design, it will still sometimes occur that every day use of our instruments will suggest points in which improvements are possible; and we shall be extremely obliged for any such suggestions.

Hartmann & Braun.

I.

Apparatus for use in reading Reflecting Instruments.

Reading Telescopes.

Large Size Telescopes, fully mounted, in well designed massive framework; absolutely without iron. Telescope with powerful objective, Steinheil eye-piece, rack and pinion adjustment to do, which is available down to 1 metre distance from the object; vertical and horizontal micrometer adjustments. Angle of elevation 50—60⁰.

Scale holder for use with wood or glass scales, adjustable for height and fitted with swivel arrangement for use with upright scale.

To facilitate accurate adjustment on to the mirror, the upper part of the instrument is adjustable by means of rack and pinion for 60—75 mm from the tripod base, which is fitted with levelling screws.

No.	TELESCOPE		Magnifies	PRICE
	Aperture	Focus		
355	30 mm	25 cm	18 times	M. **330.—**
356*)	40 mm	32 cm	24 times	M. **440.—**
357	55 mm	55 cm	30 times	M. **550.—**

If required fitted with **Euryscopic Micrometer Eye-piece** Extra M. **15.—**.

EXTRAS.

1) **Level** for use on the horizontal plane of telescope thus enabling the instrument to be used to determine the meridian, by corresponding readings on a star . . . . M. **25.—**

2) **Circular level,** fitted on the base of the telescope support M. **16.—**

3) **Diaphragms** fitted to cover, to reduce the aperture of the objective M. **12.—**

4) Steinheil **Eye-pieces,** various powers, interchangeable, each M. **15.—**

5) **Euryscopic Micrometer Eye-pieces,** Mittenzwey type, each M. **30.—**

6) **Adjustable Prisms,** Neumeyer type, for use in front of the objective, to read two or three instruments from one spot, with the requisite scale holders; made only for No. 357 . . . . each M. **150.—**

_____

*) The most convenient size.

$\frac{1}{5}$n.Gr.

No. 356.
Reading Telescope for use with Reflecting
Instruments.

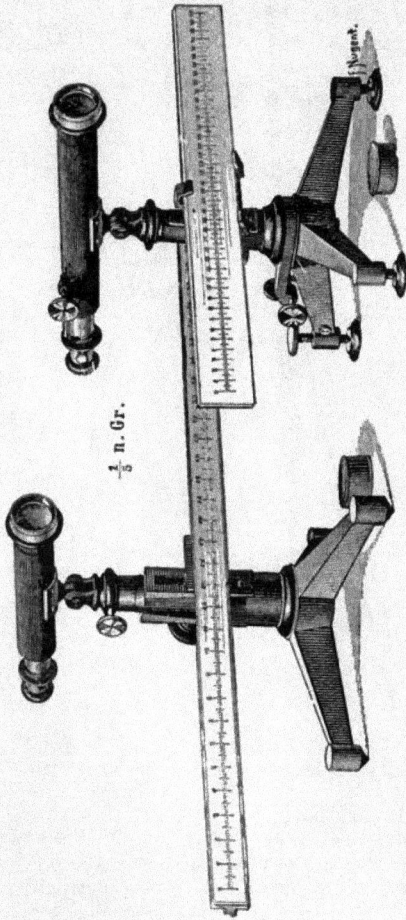

⅓ n. Gr.

No. 359.
Simple Reading Telescope.

No. **359**. **Simple Reading Telescope**, 27 mm aperture, 18 or 21 cm focus, medium power. General design resembles No. 358 but rack and pinion adjustment to eye-piece and micrometer adjustments are not provided. Telescope stand in zinc; suitable for most classes of scientific work. Price M. **80**.—

No. 358.
Small Reading Telescope.

No. **358**. **Small Reading Telescope**, 27 mm aperture, 25 cm focus; with rack adjustment to eye-piece and micrometer adjustment to the horizontal plane. Brass tripod stand with an adjusting screw for fine adjustment in the vertical plane; height of scale holder adjustable. Price M. **150**.—

The scale holders are designed for use with either glass or wood scales, in No. 359 the scales can also be used vertically.

Reading Telescopes, not mounted.

No. 350. **Reading Telescope**, aperture of objective 10 mm, focus 6 or 9 cm, with Ramsden eye-piece, equiv.-focus 1 cm, sliding adjustment to eye-piece. M. **25.**—

No. 351. **Reading Telescope**, aperture 15 mm, focus 9 or 12 cm, with Ramsden eye-piece, equiv.-focus 1 or 1.25 cm with sliding adjustment to eye-piece M. **36.**—

No. 351a. **As above,** but with rack & pinion adjustment M. **45.**—

No. 352. **Reading Telescope**, aperture 27 mm, focus 18, 21 or 25 cm, with Ramsden eye-piece equiv.-focus either 1, 1.25, 1.5 or 2 cm, with sliding adjustment to eye-piece . . . M. **50.**—

No. 352a. **As above,** but with rack & pinion adjustment M. **60.**—

No. 353. **Reading Telescope**, aperture 30 mm, focus 25 or 32 cm, with Steinheil eye-piece equiv.-focus 1.5 or 2 cm, and rack and pinion adjustment to do M. **75.**—

No. 354. **Reading Telescope**, aperture 40 mm, focus 32 or 40 cm, with Steinheil eye-piece equiv.-focus 1.5 or 2 cm; and rack and pinion adjustment to do M. **100.**—

Any of these telescopes can be fitted with **euryscopic micrometer eye-pieces** at an extra cost of M. **15.**—

$^{1}/_{16}$ n. Gr.

Stands.

No. **348. Plain Stand,** height 175 cm, in light wood, with metal sliding clamp and hinge for use with any of the above telescopes, and scale holder. Price M. **70.**—

No. **348a. As above,** but with metal sliding clamp and micrometer adjustment to telescope in both planes, with scale holder . . M. **125.**—

No. **349. Gauss Stand.** Legs in solid oak with triangular section rising bar in light wood oak veneered to match; strong wood clamping screws, size of table 35 cm. Price M. **55.**—

No. **349a. As above,** for larger instruments, size of table 50 cm. Price M. **75.**—

Arrangement for fixing reading telescopes on a cross bar. — The price varying in accordance with the mounting required, we can only quote on receipt of particulars.

Scales.

No. 360. Wood Scales of **T** section, made of well seasoned material with narrow paper strips, machine divided at our works, with mirror figures.

No. 360a. Transparent Celluloid Scales, with one edge mounted in wood bar, for use with reading lamp.

Length in cm			40	60	80	100
Wood	1 division = 2 mm*) M.		5.–	7.–	9.–	11.–
	1 „ = 1 „	„	6.–	8.–	10.–	12.–
Celluloid 1 „ = 1 „		„	9.–	12.–	16.–	21.–

*) Most suitable for use with low power telescopes.

No. 361 and **361a. Scales on plate** and **opal glass**. The glass strips for these scales are specially prepared, ground and polished true on both surfaces, with finely etched millimeter divisions and mirror figures.

Length in cm	40	60	80	100	120	140
Plate glass . M.	15.–	20.–	25.–	30.–	36.–	42.–
Opal glass. . „	16.–	21.–	27.–	35.–	45.–	55.–

We shall be pleased to supply these scales to any length or pattern, with **consecutive marking, or scaled from the centre;** also scaled for direct readings. If desired the plate glass scales can be supplied on finely ground glass, thus enabling the graduations to be illuminated from the back, or for use as direct reading.

Our opal glass scales have on account of their brilliancy been adopted by a great number of laboratories. Several hundreds are in use.

¹/₈ n. Gr.

Wood case, plain or polished, for glass scales. Price according to length. From M. **8.**— to M. **20.**—

Tubular Gas Rod, in brass, with cock and steatite burners to illuminate the scale by a number of small gas jets. Price M. **10.**—

No. 362. Lantern in sheet copper, with slot or diaphragms, adjustable convex lens, and scale-holder. Fitted either with gas, petroleum, or incandescent lamp, on wood tripod stand. Price M. **85.**—

No. 362a. As above, but adjustable for height, with metal tripod stand. Price M. **125.**—

No. 363. Mirror Distance Indicator. This consists of maple rod divided in centimetres, which can be extended to 4 metres, having a fixed contact for the scale end and an ivory contact with micrometer adjustment for use on the mirror; a suitable stand provides a means of fixing the extended rod in any desired position. Price M. **200.**—, without stand M. **160.**—

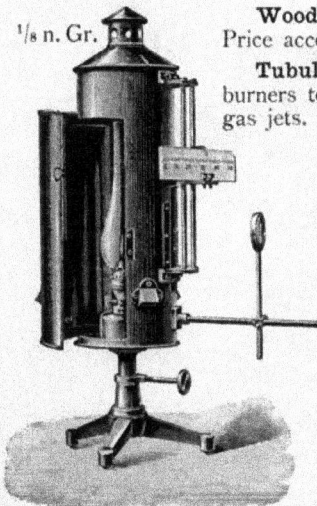

No. 362a.

II.

Galvanometers.

In addition to certain special types, the following galvanometers (for small currents) may be divided into three classes; namely, galvanometers with **bell** (thimble) **magnet**; with **moveable coil**; and with **astatic magnets**.

In designing these galvanometers the chief points we have kept before us are, that they shall be portable, easily and quickly adjusted, practically dead-beat and of the highest attainable degree of sensitiveness; the sensitiveness given for each instrument shows that we have carried out this last stipulation in a very high degree.

In galvanometers of the **first class** we have attained this by flattening the copper damper, thus enabling the winding to be brought close up to the bell magnet, which we have also made smaller than formerly and are thus enabled to increase the number of turns in the coil. Great care is taken in the manufacture of the modified type of Siemens bell magnet; they will carry 50—75 times their own weight, and therefore cause a powerful damping action although they have ample play in the damping case.

The **second class** is noticeable for their powerful, homogeneous magnetic field, thus rendering them unaffected in the proximity of other magnetic fields, and for their extremely even scale through wide limits.

In the **third class** the suspension fibres are, as far as practicable, quarz filaments.

In all the various designs great care has been taken to protect the moving parts from air currents. The mirror galvanometers are mounted with true plane mirrors, accurately surfaced, and silvered at back, which are clamped as lightly as possible in three claws; this method of suspension entirely obviates the annoyance caused by the deformation of mirror. The mirror framework is adjustable as regards the magnet, the instruments can therefore be read at any azimuth. The front of mirror case is fitted with a true plane glass aperture canted slightly forward, thereby obviating reflections.

No. **366. Simple dead-beat Galvanometer with pointer,** powerfully damped bell magnet on cocoon fibre suspension. Coils wound with two separate wires in parallel. All parts are plainly visible within the round wooden case which can be turned on the metal tripod stand. A safety catch is provided to control the swing of the pointer and for travelling purposes.

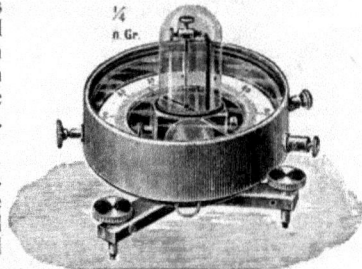

¼
n. Gr.

Made in two patterns:

I. **Low Resistance.** 2—5 ohms. The coil windings attached to three terminals so that each coil can be used separately, or both in series as well as differentially.

Sensitiveness: 1° on scale = 0.00005 ampère.   Price M. **90.**—

II. **High Resistance.** With any desired resistance up to 1000 ohms. The coil windings attached to four terminals, so that the instrument can be used as differential galvanometer or the windings can be coupled in parallel thus reducing the resistance to ¼.

Sensitiveness at 1000 ohms: 1° on scale = 0.000005 ampère.
Price M. **100.**—

No. 367.
Dead-beat Differential-Galvanometer.

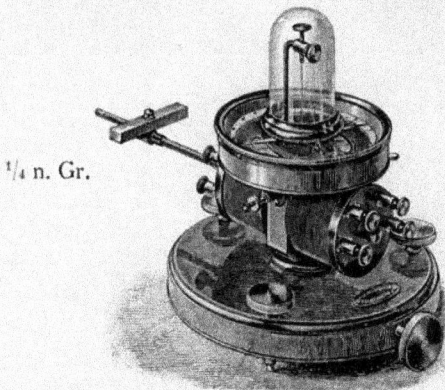

¹/₄ n. Gr.

No. **367. Dead-beat Differential-Galvanometer** with index pointer reading on German silver engraved scale with mirror underneath to obviate parallax errors. Bell magnet on raw silk fibre suspension working in a shallow but very powerful copper damper; means are provided to control the swing of the pointer, and to render the instrument portable. A circular level is provided for adjustment and the galvanometer is mounted to turn and be clamped in any position on the circular wood base which is fitted with levelling screws. The coils are removeable and interchangeable without dismounting the instrument and are adjustable to some extent as regards the magnet, thus affording a means of modifying the sensitiveness, or for accurate adjustment when used as a differential galvanometer; they are double wound with two wires of the same resistance, which can therefore be varied from one to sixteen times.

An adjustable directive magnet can be used on one side to render the instrument astatic and so increase the sensitiveness.

Sensitiveness, with the coils usually supplied of about 100 ohms; 1° on scale = 0.000008 amp.; if astatic = 0.000002 amp. Price M. **200.**—

Pair of coils, Voller's system, for measuring the resistance of incandescent carbon loops; with a resistance ratio of 1 : 20 (0.5 and 10 ohms) with screw arrangement to adjust the thick wire coil as regards the magnet for exact differential working. Price M. **60.**—

Coils, of various windings, interchangeable with those supplied (see next page).

No. 367a. Portable Dead-beat Galvanometer with Telescope.

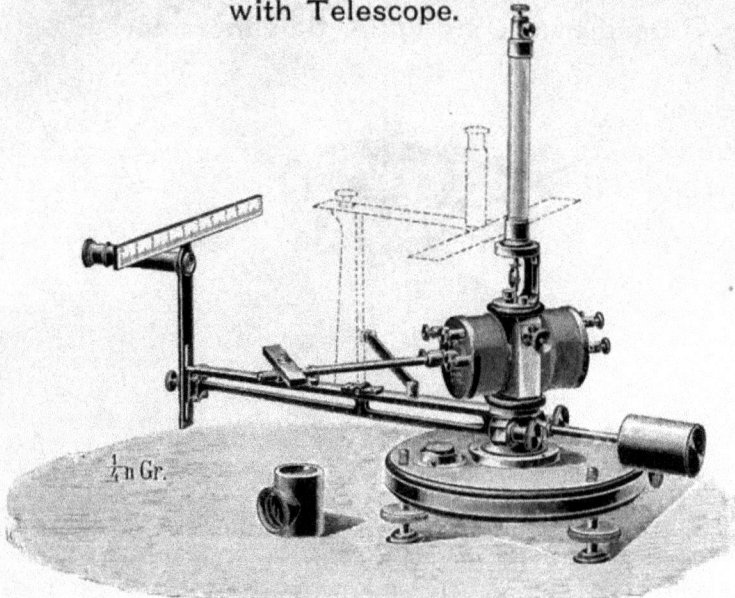

¼ n Gr.

No. **367a. Portable Dead-beat reflecting Galvanometer with Telescope.** The general arrangement of this instrument is the same as in the Differential-Galvanometer No. 367. The telescope*) has an aperture of 10 mm, focus 6 cm, thus giving ample magnifying power with great brilliancy; it is mounted on a counterbalanced arm and can be turned independantly of the galvanometer. The mirror is adjustable as regards the magnet. To enable the winding to be easily adjusted to the magnetic meridian a pole indicator to the magnet is mounted above the mirror. Stop, level, and magnet for astatic work, same as in No. 367. The instrument is suitable for either laboratory, workshop, or outdoor use and sufficiently sensitive for most galvanometric measurements.

Sensitiveness, not astatic, using the coils of about 100 ohms supplied with the instrument: 1 mm deflection on scale at 25 cm distance $= 0.0000009$ amp.; with coils of 4000 ohms $= 0,0000003$ amp.   Price M. **285.—**

Magnet arrangement for astatic work, as in No. 372. Price M. **40.—**

Coils, of various windings interchangeable for this instrument, or No. 367 of any desired resistance up to

a total resistance of 1000 ohms per pair Price M. **40.—**
„	„	„	„	2000	„	„	„	„	„ **50.—**
„	„	„	„	4000	„	„	„	„	„ **60.—**

*) If the telescope and counterbalanced arm are not required, deduct . M. **70.—**

9

No. 371 and 371a.

Dead-beat reflecting Galvanometer.

In this design we have combined all the advantages of the Siemens' bell magnet, the Wiedemann adjustable coils, and Braun's method of astatization by means of a soft iron ring, with several mechanical improvements.

By reducing the size of the bell magnet and the sectional area of the copper damper whilst retaining all its former damping power, and suspending the damper on comparatively thin wires, we have succeeded in bringing the winding so close round the poles of the magnet, that the instrument is now ten times as sensitive as formerly. The iron ring is in two parts, this arrangement allows their adjustment relatively to each other and therefore should any polarity arise in the ring, it can be easily eliminated, whilst it is easily removeable if not required.

The coils are double wound with silk covered wires, the windings can therefore be coupled up so that the resistance varies as desired from either unity to two, four, eight, or sixteen fold. The total resistance of the coils supplied with this galvanometer is about 400 ohms; they are, of course, interchangeable with similar coils of various resistances. Each coil has, in addition, some turns of thick wire.

A circular level is mounted on the tripod stand to facilitate the use of the instrument.

No. **371. Dead-beat reflecting Galvanometer. Large size.** Can be turned and clamped in any desired position, mounted on heavy metal tripod base. Movement of coil controlled by rack and pinion; the coil bar divided into millimetres.

Sensitiveness with coil coupled in series, 400 ohms resistance, and without the use of the iron ring for astatization: 1 mm deflection on the scale at 1 m distance — 0.00000008 amp.   Price M. **520.—**

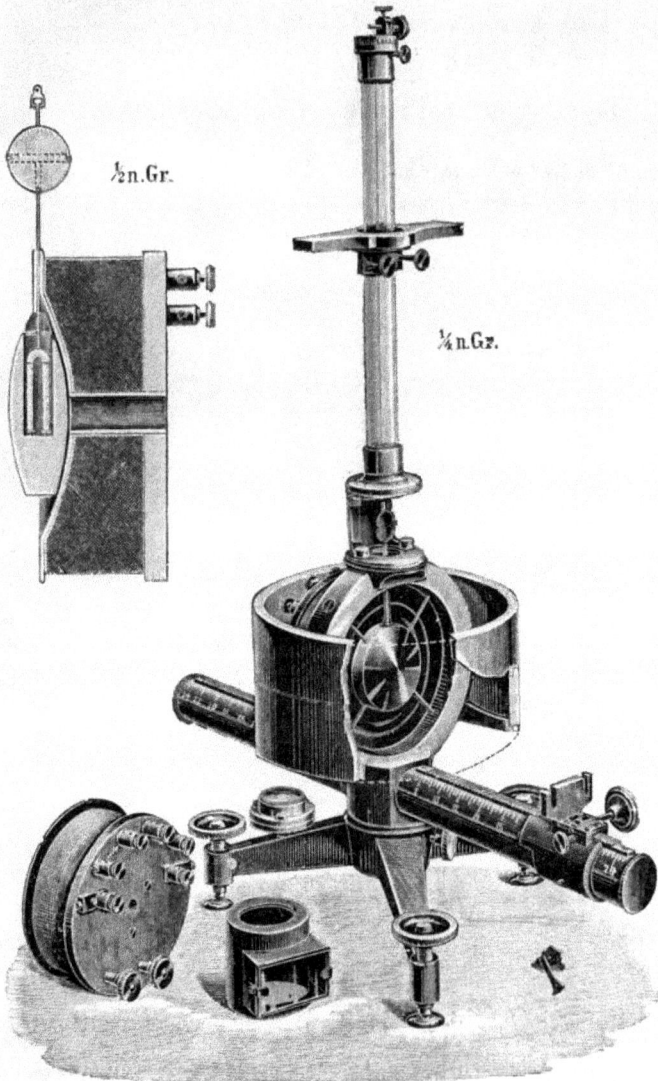

½n.Gr.

¼n.Gr.

No. 371.
Dead-beat reflecting Galvanometer. Large size.

No. 371a. Simple Dead-beat reflecting Galvanometer.

⅛ n. Gr

No. **371a**. **Simple Dead-beat reflecting Galvanometer**, resembling No. 371 throughout, but simplified in the details; i. e. with wood base, sliding adjustment to coils, and the coil bar divided in half centimetres. Sensitiveness, approximately the same as No. 371. Price M. **390.—**

Coils, of various windings, double wound, interchangeable for this instrument or No. 371, with a total resistance of about 1000, 2000, or 5000 ohms . . . . per pair M. **70.—**, **80.—** or M. **90.—** as above, but of extremely low resistance, for thermal currents M. **55.—** or with winding insulated by gutta percha for frictional electricity M. **50.—** Prices for any desired winding on application.

Control Magnet (Hauy's rod) for astatization, mounted on suspension tube with requisite adjustments for rotation and extension. Preis M. **25.—**

In both of these instruments the **Wiedemann ring-magnet** with suitable damper will be supplied instead of the Siemens' bell magnet, without extra cost, if desired.
A very convenient arrangement, enabling either type of magnet to be used and quickly changed, can be supplied at an extra cost of . . . . M. **50.—**

No. 373.
Lecture-Room Galvanometer.

$4\frac{1}{2}$ n. Gr.

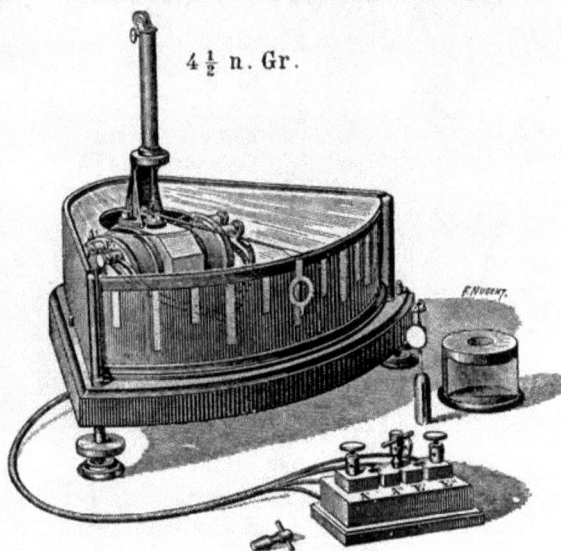

No. **373. Lecture-Room Galvanometer,** with very light pointer visible at long distances against a scale of white lines on a black ground.

Bell magnet on raw silk fibre suspension, strongly damped by suspension in narrow copper core. Coils with a total resistance of about 10 ohms, are slightly adjustable as regards the magnet; all working parts under glass; the various windings on coils are brought out on to a separate plug switch and can thereby be coupled either in series, in parallel or differentially. The scale can always be arranged to face the audience, as the instrument can be turned on its stand and the pointer*) easily adjusted as regards the magnet.

Sensitiveness: 1 Space (= 5 degrees) = 0.00015 amp. Price M. **230.**—

Thimble magnet, with true plane mirror and mirror case, fitted with accurately surfaced glass front; for use for exact measurements.

Sensitiveness: 1 mm deflection on scale 1 m distant = 0.0000005 amp.
Price M. **40.**—

Coils, of various windings, for use in lieu of those supplied, with resistance of about 1, 400, 1000, 2000, or 4000 ohms.
Price per pair M. **30.**—, **35.**—, **40.**—, **50.**— or M. **60.**—

*) When adjusting the pointer as regards the magnet, the milled disc above same should be held between two fingers of one hand and the carriage carrying the pointer between two fingers of the other hand; it is of course adviseable not to put any strain on the silk fibre whilst this is being done.

Deat-beat Galvanometers with suspended moveable coil.

Deprez-d'Arsonval System (arrangement of details patented).

No. 535a. $^1/_4$ n. Gr.

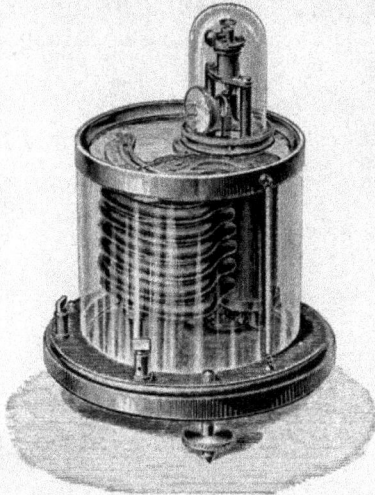

These instruments have been improved against the original type of **suspended coil galvanometers** of Deprez and d'Arsonval. A number of powerful magnets with common pole pieces produce a very strong and at the same time homogeneous field throughout the entire range of the moveable coil. The fram upon which the moveable coil is wound, is of aluminium and works in an extremely narrow clearance space, thus attaining perfect aperiodicity, independent of the outer circuit, and an extreme sensitiveness. The coil is suspended either between very thin wires which act in opposition to the current, and are taking at the same time the current to the coil, or — convenient for certain purposes — on a raw silk fibre in which case a two-way spring acts as counter force, or finally on a quarz fibre. In the two latter cases the current is taken to the coil through excessively thin strips of silver foil which have only an inappreciable effect on its movement. Owing to their extremely powerful magnetic field these galvanometers are unaffected by the earth magnetism or by other extraneous magnetic influences and can therefore be used in proximity to dynamos. They are also less sensible against vibrations.[*] Their proportional deflections render them serviceable for a great many applications.

[*] If the pointer galvanometer No. 535 as well as even the mirror galvanometer No. 535a ar to be used on places which are relatively contineously shaked (workshops, engine houses &c.) the coil can be fitted with a glycerin damper.   Price M. 35.—

No. 535. Suspended moveable coil Galvanometer, easily set up and adjusted by levelling screws in base and circular level; is safely portable, as the glass case is firmly fastened and a travelling stop is provided which, in one movement, slackens the fibre suspension and clamps the moveable coil; this stop is under the small glass bell, which is mounted on the outside case by a bayonet joint, to render it accessible. The galvanometer is read by means of a pointer moving over an arc divided in degrees; a mirror to eliminate parallax errors is mounted underneath same.

Sensitiveness: resistance of coil usually supplied about 50 ohms. 1° deflection $= 0.000002$ amp. Price M. **200.—**

No. 535a. A similar instrument as above, but **arranged as reflecting galvanometer.** The true plane reflecting mirror or, if preferred, a light concave mirror is frictionally held in an aluminium framework mounted on the moveable coil. The glass bell has a sight aperture in it covered by an accurately surfaced glass which is canted slightly forward.

Sensitiveness: resistance of coil about 2000 ohms. 1 mm deflection on scale 1 m distant $= 0.00000002$ amp. Price M. **230.—**

Mirror casing with small **reading telescope** the axis of which forms with the plane of mirror an angle of about 45° so that the deflection of same may be observed at any mark or scale properly placed at any desired distance. (See left hand figure below.) Price M. **50.—**
Price instead of ordinary mirror case M. **40.—**

No. 535 or 535a if wound as **Differential-Galvanometer** Extra M. **40.—**
No. 535a if constructed as **Ballistic Galvanometer** . „ „ **20.—**
A second winding to the latter to steady same . . „ „ **60.—**

to No. **535a.**

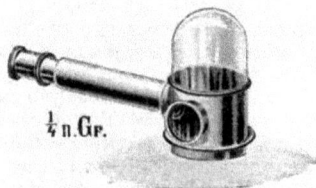

$\frac{1}{4}$ n. **Gr.**

No. **536.**

$^{1}/_{4}$ n. Gr.

No. 536. Simple moveable coil Galvanometer, with coil on steel pivots in jewelled centres, with travelling stop, very portable, and ready for use without further adjustments. Pointer reading on paper scale divided in degrees. Brass case and wood base.

Sensitiveness: coil of about 50 ohms. $1^\circ = 0.000025$ amp. Price M. **135.—**

No. 536a. Similar instrument as above, but with only three magnets instead of six; is not quite as high, and therefore still more portable.

Sensitiveness: coil of about 10 ohms. $1^\circ = 0.00005$ amp. Price M. **100.—**

For instruments of similar design, but for use with large currents, see No. 530, p. 28, and No. 601, p. 92.

No. 368.
Kohlrausch Mirror Galvanometer.

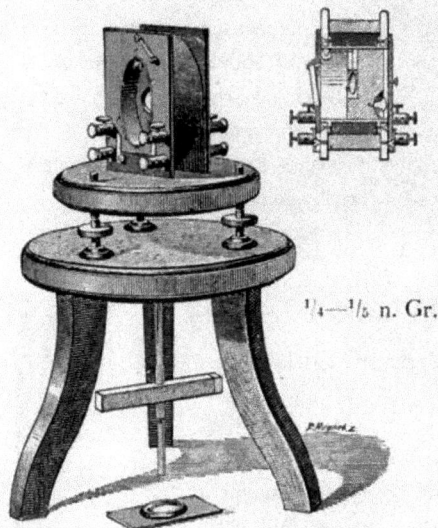

$^1/_4$—$^1/_5$ n. Gr.

No. **368. Kohlrausch Mirror Galvanometer**, with undivided oval coil in the interior of which the magnet, a steel mirror, is suspended on a silk fibre. The damping can be adjusted within wide limits by moving the copper core. The coil has both thick and thin windings, each of which is composed of two wires of equal resistance wound side by side; the instrument can therefore be used as differential galvanometer. To eliminate thermo currents these windings end in copper terminals.

The various ways in which the circuits may be connected and the use of an adjustable control magnet which is fixed underneath the stand enable the sensitiveness to be varied as desired.

Sensitiveness with a resistance of about 50 ohms and not astatic: 1 mm deflection on scale 1 m distant $= 0.000\,0004$ amp.

With Stand and control magnet . . . . Price M. **200.—**
Without Stand . . . . „ „ **180.—**

If the **steel mirror** is polished on both sides so that readings can be taken from either East or West . . . . Extra M. **25.—**

Guard Ring, in soft swedish iron, rendering the galvanometer more independent of iron in the neighbourhood or other influences, and at the same time considerably increasing the sensitiveness.   Price M. **25.—**

No. 369. Rosenthal Micro-Galvanometer. In lieu of the horseshoe shaped magnet with hooked ends and vertical suspension as suggested by Rosenthal, we employ in this instrument a somewhat **Z** shaped magnet with mirror, which is rendered astatic by the Ferrini process in which similar poles are produced at both ends of the magnet. The poles project within the coils. The Rosenthal arrangement gives extreme sensitiveness for a minimum amount of wire employed, and, in combination with the Töpler air damper, furnishes a galvanometer which is most suitable for thermo-currents or for physiological purposes.   Price M. **300.—**

No. 370.
Astatic Mirror Galvano-
meter.

No. 370. **Astatic Mirror Galvanometer** with compound magnets made up of thin laminae and each surrounded by a coil. Adjustable copper sheath to control damping action, adjustable mirror; mounted on marble base to turn in metal tripod stand. All parts are easily accessible on removal of the circular glass cover.

Sensitiveness: with a total resistance of about 10 ohms and with a 10 seconds oscillation $= 0,000\,000\,15$ amp. for a 1 mm deflection on scale 1 m distant.   Price M. **280.—**

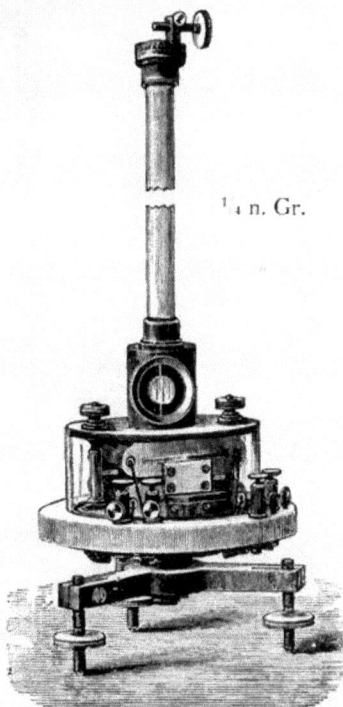

$^1/_4$ n. Gr.

No. 370a.

Astatic Mirror Galvanometer for ballistic work.

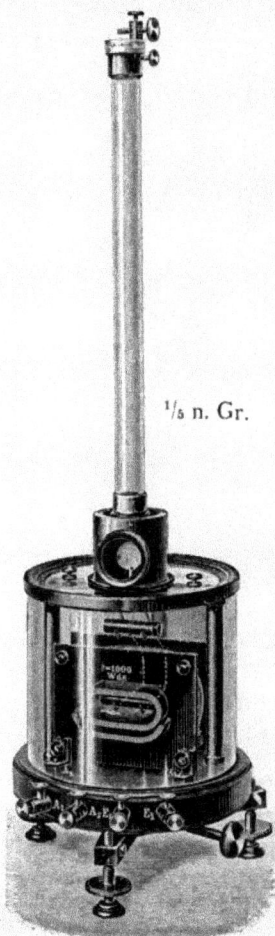

¹/₅ n. Gr.

No. 370a. **Astatic Mirror Galvanometer** for ballistic work with interchangeable tubular magnets, 40 mm long, and adjustable mirror; the magnet carriage is of aluminium and can be introduced into the coil from the outside. The coil is double wound with two similar wires side by side and another coil of any other winding can be used in its place. The damping action is controlled by an adjustable copper sheath, which can be entirely removed, if desired. All parts are easily accessible after the strong glass cover is removed. The complete galvanometer can be turned in its metal stand.

Sensitiveness with a resistance of 50 ohms and a 15 seconds oscillation: 1 mm deflection on scale 1 m distant = 0,00000002 amp.   Price M. **350.—**

Interchangeable Coils, of any desired winding up to 1000 ohms.

Price M. **50.—**

No. 370b.
Large Astatic Mirror Galvanometer.
Weber type.

No. 370b.

Astatic Mirror Galvanometer Weber type, with all modern improvements. Tubular magnets easily inserted and interchangeable, 120 mm long, mirror adjustable. The coil is double wound, but not interchangeable; adjustable copper damper. This instrument, which is of extremely good design, is in all other respects similar to No. 370a, with the exception that the glass cover is oval in shape and that the working parts are mounted direct on the base, and cannot therefore be turned in same.

It is identical with the Galvanometer No. 427 p. 74 for use with the Weber Earth-Inductors.

Price M. **400.**—

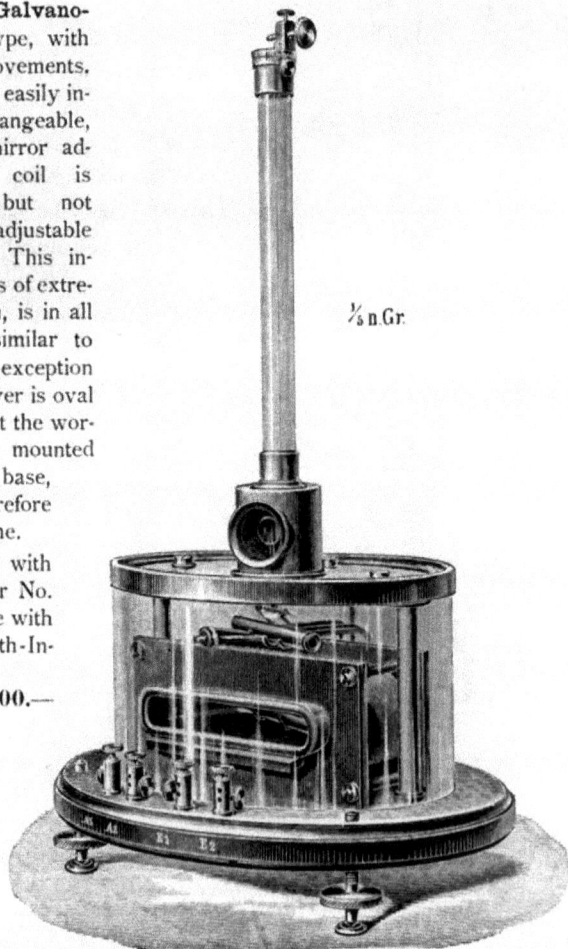

⅓ n. Gr.

No. 372.
Astatic Dead-beat Mirror Galvanometer.

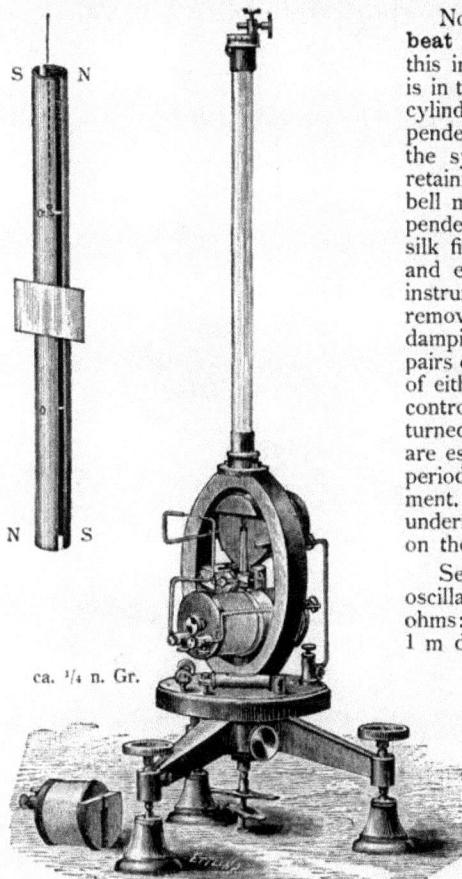

S N

N S

ca. ¼ n. Gr.

No. 372. Astatic and Dead-beat Mirror Galvanometer. In this instrument the magnet-system is in the form of two hollow semi-cylinders coupled together and suspended vertically, thus rendering the system highly astatic, whilst retaining all the advantages of the bell magnet. The system is suspended on either quarz or raw silk fibre, the mirror is adjustable and encased in the centre of the instrument, the casing can be removed if required. The copper damping action is adjustable. Two pairs of coils with a total resistance of either 5000 or 10000 ohms; two control magnets which can be turned and adjusted as required, are especially useful to control the period of oscillation of the instrument. Ebonite pillars are provided underneath the levelling screws on the base.

Sensitiveness: with 10 seconds oscillation and resistance of 5000 ohms: 1 mm deflection on scale 1 m distant = 0.000 000 0025 amp.
Price M. **550.—**

Interchangeable coils,*) two pairs of any desired lower resistance.
Price M. **150.—**

Extremely light magnet-system, on the same principle as above but using very thin rods in lieu of the semi tubular magnets, for use for the highest sensitiveness.
Price M. **30.—**

Instead of the magnet-system above described, Lord Kelvin's pattern will be supplied, if desired, without extra cost.

---

*) If required these coils can be supplied wound with wire gradually increasing in size from the centre outwards.

No. 372a.

Astatic Galvanometer for cable-work with reading telescope.

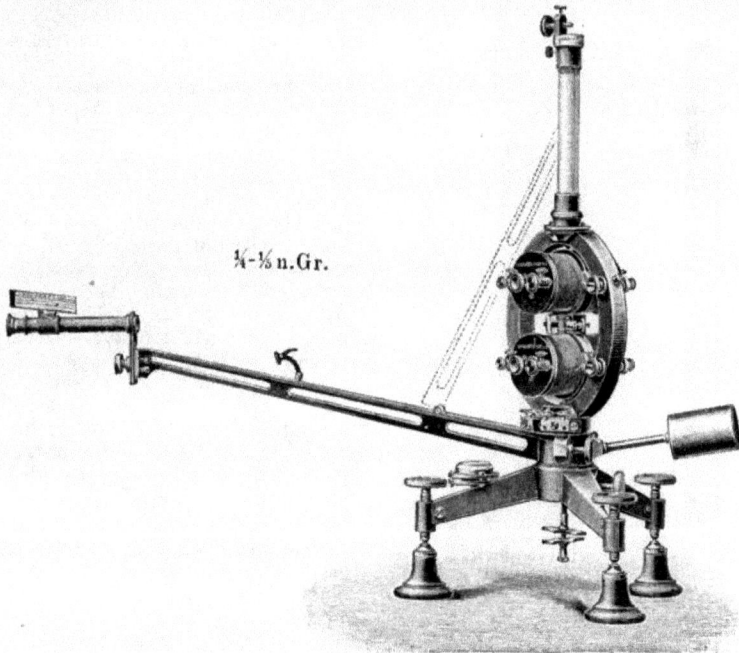

¼-⅛ n.Gr.

No. 372a. Portable Astatic Galvanometer for cable-work.
Similar in design to No. 372 but with short silk fibre suspension and fitted with telescope carried on a balanced arm which can be turned round the stand. The small telescope is mounted about 50 cm away from the mirror and has an adjustable scale attached. To render the instrument as portable as possible the telescope arm is hinged to enable it to be folded upwards as shown by the dotted lines; and a travelling stop is fitted to the magnet system. A circular level facilitates quick adjustment and the instrument is extremely serviceable wherever portability is required as for instance for use in a cable testing cart.

Sensitiveness: approximately the same as No. 372. Price M. **650.**—

No. 378.

Unifilar - Electrodynamometer for small currents.

Kohlrausch type.

No. 378. The Mirror reading Electrodynamometer by Kohlrausch is especially intended for use instead of galvanometers when measuring the resistance of electrolytes, for which purpose it is more portable and easier to erect. Galvanometers with bifilar suspension are not sufficiently sensitive for many purposes, if the fibre suspensions are fairly far apart, whilst if brought close together they are awkward to suspend properly and are as a rule insecurely mounted. In this instrument the inner moveable coil is hung on one wire only which serves at the same time to carry the current to the coil, the other lead underneath being either an extremely fine silver foil strip, or, if for alternating current, a platinised platinum electrode which dips into a small glass receiver filled with dilute sulphuric acid (15%) and mounted underneath. This electrode works between a U shaped piece of platinum foil similarly treated and is consequently of service in damping the instrument.

The suspension wire is fastened at the top to a torsion circle which is fitted with a vernier; by means of a micrometric adjustment, the plane of winding in the moveable coil can be exactly adjusted vertically to that in the fixed outer coil.

The internal moveable coil is of ivory and the thin oval true plane mirror is fixed thereto and also a counterweight for same in the form of a brake; a bundle of soft iron wires can be inserted into the ivory coil. The external fixed coil, which is somewhat oval in section and closely surrounds the inner flat coil is mounted on a brass case, cut across to eliminate induction, with ebonite flanges; it is set up in two halves and the winding on each is arranged for use either separately, parallel or in series.

To facilitate fixing the outer coil winding in the magnetic meridian, the whole instrument can be revolved in its tripod stand.

Sensitiveness or constant, with coils of 120 ohms total resistance in series, and with the iron wire bundle inserted: 1 mm deflection on scale 1 m distant $= 0.00007$ amp. approx. $(\therefore J = 0.00007 \times \sqrt{\alpha})$.

Telephonic currents can be demonstrated with this instrument.

Price M. **370.**—

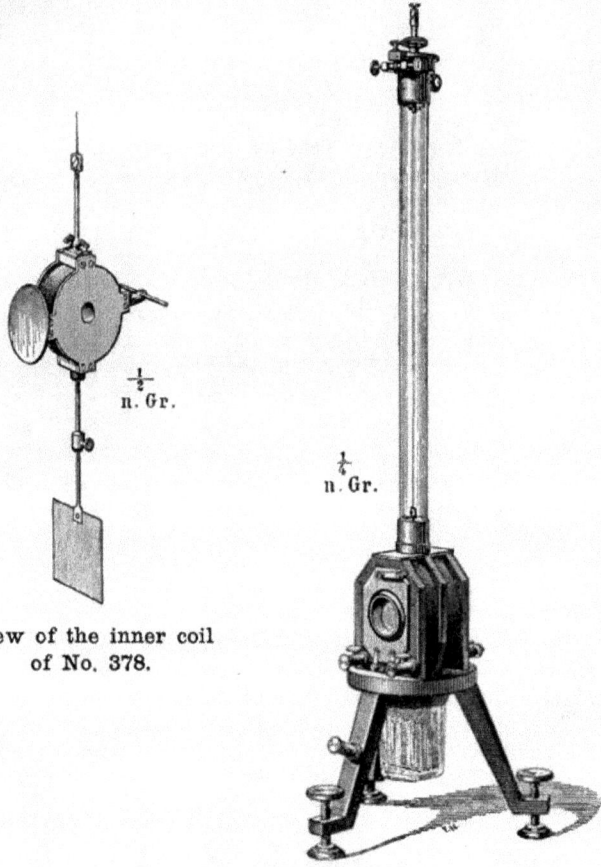

$\frac{1}{2}$
n. Gr.

View of the inner coil
of No. 378.

$\frac{1}{6}$
n. Gr.

No. 378.
Unifilar-Electrodynamometer
Kohlrausch type.

(Alternating-current Galvanometer.)

No. 374.
Tangent Galvanometer for absolute measurements,
Kohlrausch type.

No. 374.

½ n. Gr.

No. 374. Tangent Galvanometer for absolute measurements with non-metallic magnetometer. Kohlrausch type. To eliminate any appreciable local magnetic effects all unnecessary metal work is removed. The current passes through an accurately turned ring of electrolytic copper. The magnetometer (see No. 415 p. 72 & 73) has a magnet fastened to the back of the mirror, which acts also as an air-damper; it can be removed and a compass with pointer (see No. 375) used instead.
Price M. 295.—

The table on which the Magnetometer is mounted is of such a size that, according to Helmholtz (Gaugain) it can be placed about $\frac{\pi}{7}$ out of the winding-plane.

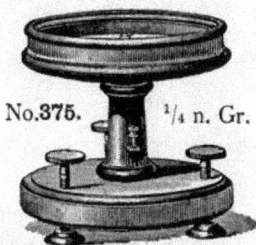

No. 375.   ¼ n. Gr.

No. 374a. As above, but with magnetometer No. 415a (see p. 72 & 73) in which the mirror can be adjusted as regards the magnet which swings in a copper damper. Price M. 355.—

No. 375. Simple Compass with small steel magnet working in jewelled cap on needle point, copper damper and check stop. Pointer reading on plate glass scale either silvered or blackened and divided in degrees; for use instead of the magnetometer on the Tangent Galvanometer No. 374 when measuring larger currents. Price M. 60.—

No. 376a.
Tangent Galvanometer for technical use.

No. 376. Simple Tangent Galvanometer with one turn of very thick wire mounted to rotate on a massive stand. Needle galvanometer with scale divided on cardboard and with inlaid mirrors to facilitate accurate reading of the thin aluminium pointer; magnet fitted with copper damper and check stop.

Price M. **210.—**

Range from 0.1 to 25 amp. An arrangement can be supplied to increase the range of the instrument up to 100 amp. without using shunts, by withdrawing the needle galvanometer a certain measureable distance from the plane of the ring.

Extra M. **40.—**

⅕ n. Gr.

No. **376a. As above,** but with a second concentric ring, smaller in diameter and wound with 2×300 turns of fine silk covered wire, the surface of which can be measured. Price M. **300.—**

Using these windings in series or in parallel small currents down to 3 milliampères can be measured and also potentials determined — using suitable resistances if necessary (see No. 408 page 43).

Voltameters.

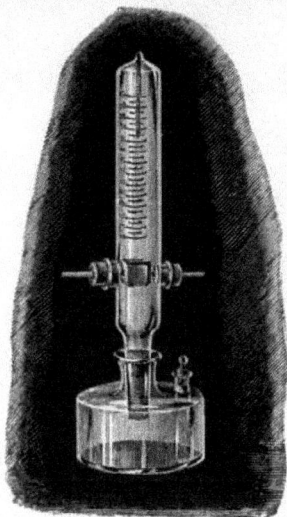

No. 385.

No. 385. Kohlrausch Water-Voltameter. This is for large currents up to 30 ampères, with thermometer sealed into the top of the gas vessel. It is especially useful when there is no accurate chemical balance at hand for weighing silver or copper deposits. Readings with this instrument occupy but little time and it is exceedingly simple and convenient.

Price: without platinum electrodes M 35.—.

Platinum electrodes: by weight and at market price, from M. 20.— to M. 30.—.

¹/₅ n. Gr.

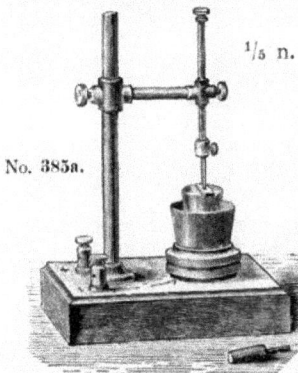

No. 385a.

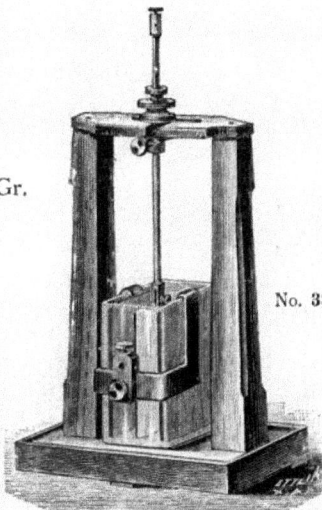

No. 385b.

No. 385a. Silver-Voltameter, with large rod or bowl shaped silver anode and stout platinum cup, for a current of 0.25 ampère. Price: without platinum cup . . . . M. 60.—
Platinum cup, by weight and at market price, from M. 40.— to M. 60.—

No. 385b. Copper Voltameter with several convenient arrangements to facilitate working; active surface of electrode about 200 sq cm, for about 5 ampères. Price, without platinum electrode . . . . . . M. 70.—
Platinum electrode, by weight and at market price from
M. 50.— to M. 100.—

III.
Direct reading Ampère-, Volt- and Wattmeters.

Direct reading instruments offer considerable advantages, compared with those in which the readings are determined by turning a deflected magnet, or coil carrying the current, back to its zero position. Firstly, because greater accuracy in the readings can be obtained as there is only one pointer to be read instead of two as in zero instruments; secondly, because no manipulation, such as turning a torsion spring is necessary, and readings can therefore be taken some distance away from the instrument and without any loss of time; lastly, because direct reading instruments can be used to measure a varying or pulsating current, and will indicate the exact current passing at any moment; provided that they are sufficiently damped; a point which has received special attention in the following instruments.

Special care has been taken in the design of these instruments to render them portable and easy to use; for these purposes they are provided with circular levels and fitted with a convenient safety stop to the moving parts for travelling purposes.

Each half of the torsion springs used in these instruments is wound in opposite direction, thereby eliminating any temperature variations which could otherwise affect the accuracy of the readings.

Direct reading Galvanometer for current and potential (patent).

No. 530.

¹/₄

n. Gr.

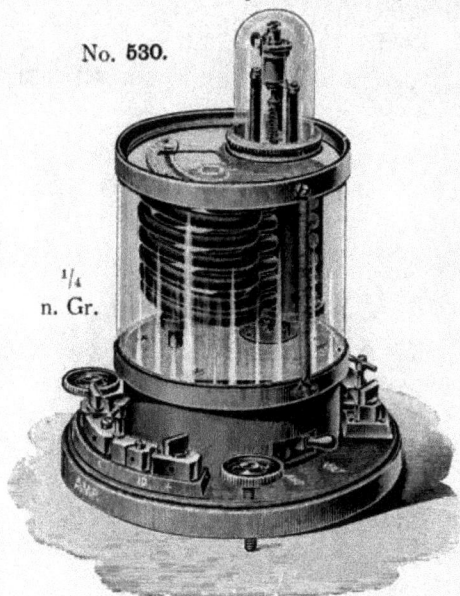

This **moveable coil direct reading dead-beat Galvanometer,** works on the same principle as No. 535 and therefore possesses similar advantages when used for larger currents. The magnetic system employed, forming an almost closed circuit, and the treatment of the magnets on a similar method to that of Barus-Strouhal, ensure that the readings are practically constant and unaffected by temperature variations. The scale is almost exactly proportional throughout, and, the pointer working practically dead-beat, renders this galvanometer extremely suitable for use as a standard or control instrument. It reads direct from one to one hundredfold or higher. Made in three types (with 3 magnets, not 6 as shown).

No. **530. Direct reading Galvanometer** with moveable coil, to read in volts and ampères; with series and shunt plug resistances arranged in the base of instrument; reading from 0.1 to 15 amp. and 0.1 to 150 volts. A suitable switch contact which serves also as commutator for current and potential measurements, enables readings of either to be quickly taken. Price M. **325.—**

No. **530a. As above;** but as ampèremeter only; with any desired scale not exceeding 15 ampères. Price M. **180.—**

No. **530b. As above;** but as voltmeter only; with any desired scale not exceeding 300 volts. Price M. **180.—**

Shunt resistance ⎱ for use between the terminals ⎰ Price M. **45.—**
Series resistance ⎰ to increase the range tenfold ⎱ „ „ **105.—**

Other shunt and series resistances can be supplied, if required.

Direct reading Electrodynamometer and Wattmeter (patent).

The principle on which these **direct reading electrodynamic instruments** are constructed is the action of a suspended moveable double solenoid with similar in- and external poles NS—SN on a conaxial, ringshaped fixed solenoid. As compared with other types of electrodynamic instruments it has the advantage of an approximately equal scale throughout its range, as any movement of the suspended solenoid in regard to the fixed solenoid does not appreciably alter their effect. The instrument is perfectly damped by an aluminium disc connected to the moveable solenoids and working beetween a powerful magnet placed in the base of the instrument, a thick iron plate above it shielding the moveable coils from its effect. The instruments can be used for measuring **alternating currents,** without alteration, if non-inductive series resistances are employed.

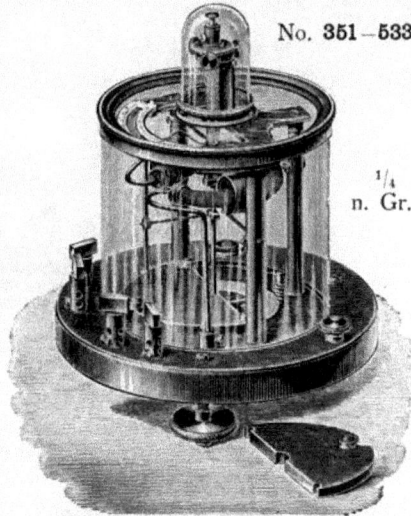

No. 351—533.

¹/₄ n. Gr.

Made in three types:

No. **531. Electrodynamometer** for large currents with any desired scale not exceeding 25 ampères.   Price M. **340.—**

No. **532. Electrodynamometer for small currents** (available also as voltmeter up to 150 volts).   Price M. **340.—**

No. **533. Wattmeter;** with any desired scale not exceeding 1500 watts with 15 amps. current.   Price M. **340.—**

Commutator to extend the scale lower.   „   „   **30.—**

If arranged as reflecting instrument . . . . . . Extra „   **50.—**

„   „   astatic . . . . „   „   **75.—**

Shunt and **series resistances;** according to agreement.

IV.
Copies of the Standards of the Imperial German Physico-Technical-Laboratory.

In reference to the following Standard copies of cells, resistances and compensation apparatus, we beg to state that, if desired, we supply, at own cost, the official Certificate from the above laboratory for each instrument.

No. **551. Standard Clark cell.** The mercury electrode formed by amalgamated platinum foil is sealed in a porous cell with the mercurous sulphate paste, thus rendering the cell portable. A thermometer projects into the cell with an external scale for easy reading. The glass is protected by a metal casing. E. M. F. — 1.438 Volt (at 15°C.). Price M. **30.**—

¹/₅ n. Gr.   No. **554.**

No. 553.   No. 552.

Standard resistances from 0,1. 1. 10, &c. up to 100000 ohms, wound with either Manganin or Constantan, and therefore requiring no correction for temperature; with nickel plated electrodes for insertion in mercury cups. The resistance coils are accurately calibrated and silver soldered to the electrodes; they are mounted in perforated metal cases and can therefore be used in paraffine baths for measurements at constant temperatures.

No. 552.		I	II	III	IV	V	VI	VII
	Ohms	0.1	1	10	100	1000	10000	100000
Strongest current admissible in petroleum bath	Amp.	5	1	0.3	0.1	0.03	0.01	0.003
	Price M.	**50**	**45**	**45**	**45**	**45**	**70**	**100**

No. 553. **Standard Resistance of 0.01 ohm** can also be used as branch resistance for accurate current measurements up to 60 ampères; similar design to those previously described but made from thin strip instead of wire; fitted with two terminals at the very ends of the resistance for potential measurements . . . . Price M. **60.**—

No. **554.** **Standard Resistance of 0.001 ohm,** for use as branch resistance for currents up to 250 ampères. The electrodes are fitted with conical plugs on which the accurately fitted cable connectors can be mounted; these connectors are bored and prepared to be soldered to the cables; a short circuiting bar is provided to cut out the resistance, and the terminals for the potential leads are mounted at the ends of the resistance which is of sheet manganin with large cooling surface and mounted in metal case filled with petroleum. A cooling worm for use with a water supply and arrangement to stir the bath enable the temperature to be easily controlled. Price M. **220.**—

Standard Resistance for 0.0001 ohm. Price according to agreement.

ca. ¹/₅ n. Gr.

No. **555.** **Petroleum baths,** for resistances Nos. 552 and 553. The vessel brazed up from sheet copper, fitted with draw off cock and thoroughly nickelled. Double mercury cups with terminals and nickelled are provided to carry the resistance and leads.

Made for **three** or **four** resistances. Price M. **70.**— or **80.**—

Extra for a **Turbine stirrer,** to drive by band.*) M. **30.**—

In addition to the standard resistances Nos. 552—554 we can, if required, supply separate accurately adjusted resistances of any other size to carry any desired current up to 3000 ampères; made in manganin or constantan. These are mounted in lead lined wood boxes for use with petroleum. Prices according to size and requirements.

---

*) Small Electromotor to work at various low potentials for driving the stirring arrangements at prices . . . . from M. **30.**— to M. **50.**—

Compensation-Apparatus
for exact current and potential measurements.
Direct reading.

The various **Compensation Apparatus** about to be described are modelled after the standards of the German Imperial Laboratory and are designed for current and potential measurements by comparing the unknown quantity with the E. M. F. of a standard Clark cell, which is mounted in the apparatus but can be removed if required for other use. The system of measurement is a zero method and the apparatus is of the same value for all current and potential measurements as the Wheatstone bridge is for all resistance work; it is in fact next to the latter, the most important instrument in an electrical laboratory. The requisite accessories are a battery with suitable switch and a sensitive galvanometer.

The standard cell is protected from a short circuit through a ballast resistance of 100 000 ohms and as the effect of the compensation is that no current passes, no variation in its E. M. F. can occur.

The circuit is branched from the main on to the switch contacts of two decade-rheostats of 10×100 and 10×1000 ohms and a set of plug contact resistance coils, and is adjustable to 0.1 ohm thereby. The greatest care is taken in calibrating and connecting these resistances. (For particulars as to official Certificate see heading to this chapter.)

On the left hand side of the apparatus is a commutator to substitute the standard cell for the unknown E. M. F. and on the right is a current switch with intermediate contact for the above mentioned ballast resistance.

Range from 0.014—1400 volts, and with suitable branch resistances, (such as Nos. 552 and 554) from 0.1 milliampère to 1000 ampères.

No. 556.
Large Compensation Apparatus.

ca. $\frac{1}{4}$ n. Gr.

No. **556. Large Compensation Apparatus,** arranged as described on opposite page, with a main circuit of somewhat over 111 000 ohms adjustable throughout, the adjustment for temperature of the standard cell is thus included in the apparatus and the instrument is therefore direct reading.   Price M. **660.—**

Instructions
for using the large Compensation
Apparatus No. 556.

1. **Measurement of Potentials between 14 and 1400 volts:** The points between which it is desired to ascertain the difference of potential are connected to **B +** and **B −** and the Galvanometer to **GG**, with the commutator turned to the right on to **c +** and **c −**. The right hand switch contact is placed on 1×1000 and the left hand on 4×100 whilst the plugs corresponding to $53 - t$ are withdrawn from the long front row of resistances where t represents a figure read off on the thermometer of standard cell. The compensation circuit is now closed by turning the switch from x on to 100000 and the deflected galvanometer approximately brought back to zero by plugging in the horseshoe and smaller row of resistances; and for fine adjustment turning the switch on to O and altering the resistances until the galvanometer gives no deflection when the circuit is alternately made or broken. The unknown potential E is then $0.001 \times W$ volt where W is the sum of all plugs withdrawn plus the resistance of the circular rheostat contacts 10000.

2. **Measurement of Potentials under 14 volts:** a battery having an E. M. F. at least ten times as large as the one it is desired to measure, is connected to **B +** and **B −**; the unknown E. M. F. to **A +** and **A −** and the galvanometer to **GG**.

Two measurements are now necessary: the commutator is again turned first to the right, and the remaining operations are carried out as in 1 with the exception that all the plugs are withdrawn from the inner row of resistances and only those reinserted which correspond to those withdrawn on the outer row to evaluate t.

Without considering the resistances in circuit when the compensation is attained, the commutator is then turned to the left, and an approximate compensation obtained with the switch on 100000; firstly by altering the switch contacts on the circular decade rheostats, and finally — with the switch on O — by shifting the plugs in the two front rows of resistance coils, whatever resistance plug is removed in the one row being inserted in the corresponding coil in the other so that the total resistance in circuit remains unaltered. The unknown potential A is then $0.001 \times W$ volt where W is the sum of the resistances inserted between the switch rheostat contacts.

3. **Measurements of current.** These are always correlated to differences of potential by passing the unknown current through an accurate standard resistance[*]) of suitable size and determining the potential difference at its terminals as described No. 2 above. The result is read off direct without working out if the resistance employed is 10 Ohms or a $\pm$ power of 10 Ohms.

---

[*]) see Nos. 553 & 554, p. 30 & 31.

No. 557.

Simple Compensation Apparatus for exact current and potential measurements.

ca. $\frac{1}{4}$ n. Gr.

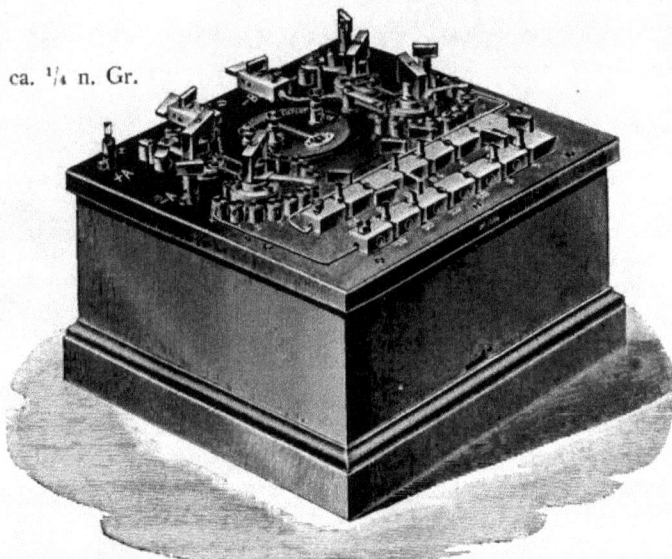

No. **557. Simple Compensation Apparatus,** arranged as described on opposite page, but with main circuit of somewhat over 100000 ohms, of which 90050 ohms is not adjustable; the results obtained, therefore, require correction for temperature.   Price M. **475.—**

This simpler instrument is used similarly to No. 556.

To measure high potentials the galvanometer and the unknown E. M. F. are connected as above; the commutator on **c +** and **c −** and compensation attained by turning the switch contacts and varying the plug resistances. If W represents the resistances in circuit between the switch contact resistances, then the unknown E. M. F. $E = \frac{100}{W} (1453 - t)$ where t again represents the temperature of the Clark cell.

For use with low potential measurements, the unknown E. M. F. is connected at **A** and a suitable battery (see opposite page) at **B** and twice compensated for; firstly the Clark cell with the commutator on **c +** and **c −** and secondly the unknown E. M. F. with the commutator on **A**. Then if W_1 and W_2 represent in each case the resistances in circuit between the switch contacts $A = \frac{W_2}{W_1} (1.453 - 0.001\ t)$ volt.

V.
Apparatus for measuring electrical resistances.

No. 389. Wheatstone-Kirchhoff Bridge, Kohlrausch roller type. In order to obtain all the advantages of a long branch circuit without the inconvenient length, otherwise necessary, the wire in this instrument is mounted in ten turns on a white marble roller in which the circumference is divided, on a German silver ring, into 100 parts. A small friction-contact roller serves as moveable contact and indicates the turns in circuit, and the constant connections are formed through brush contacts of twenty wires. The base of the instrument contains the comparative resistances of 1, 10, 100, 1000, 10000 ohms, — with anti-inductive winding, Chaperon system —, which, through plugging can be inserted on either side of the measuring wire. Price M. 250.—

Cover for above in wood with glass insets, to shield the measuring wire from sudden changes of temperature. Price M. 15.—

If desired, and at an extra cost, **additional resistances** can be supplied which by means of plugs can be connected to both sides of the measuring wire, of the same or $4^1/_2$ times the resistance of the latter; or if preferred, with a **plug arrangement** to connect these resistances to either one or the other side of the measuring wire or to make them interchangeable.

No. 389a. Wheatstone-Kirchhoff Bridge with stretched wire 1 metre long and in which the resistance can be increased three fold by connecting similarly treated wires to the two ends of the measuring wire. The sliding contact carriage is well guided and the frictional contact can be removed; they work over a scale accurately divided into millimetres. Without comparative resistances. Price M. 140.—

A second contact carriage and frictional contact for calibrating the measuring wire, or for use of bridge as du Bois-Reymond current-compensator, or as Thomson (Lord Kelvin) double-bridge. Price M. 25.—

Extra set of divisions for direct reading of resistance, referring to the measuring wire only and not to the additional resistances. Price M. 10.—

No. 390. Induction apparatus, Kohlrausch type, for generating alternating currents. It is the simplest way, when measuring the resistances of electrolytes. to eliminate polarisation by using alternating currents. This induction coil, which requires 3 Daniell cells to work it, is designed for this purpose. It has massive iron core and mercury contact breaker, whilst the secondary circuit is wound in two parts which can, by means of plugs, be coupled together in any desired manner. Price M. 140.—

No. 391. Universal Bridge. Kohlrausch type, a very simplified design of instruments Nos. 389 and 390 (see page 60). Price M. 130.—

No. 392. Vessels with **platinum electrodes,** Kohlrausch type; for determining the resistance of electrolytes. Two vessels connected by U tube answer this purpose; the platinised platinum electrodes have an area of about 10 sq. cm. A set of three of these compound vessels with connecting tubes of different diameters mounted on wire stands and with one pair of platinum electrodes, interchangeable throughout the set; the price varying according to the weight and market value of the platinum from M. 70.— to M. 80.—

No. 392a. As above, Arrhenius type, in which the glass vessel is cylindrical and mounted on glass base, the two platinum electrodes are superposed and the distance between them can be varied as required. Price M. 40.— to M. 50.—

Telephones specially constructed for measuring see page 60. Price M. 15.— & M. 20.—

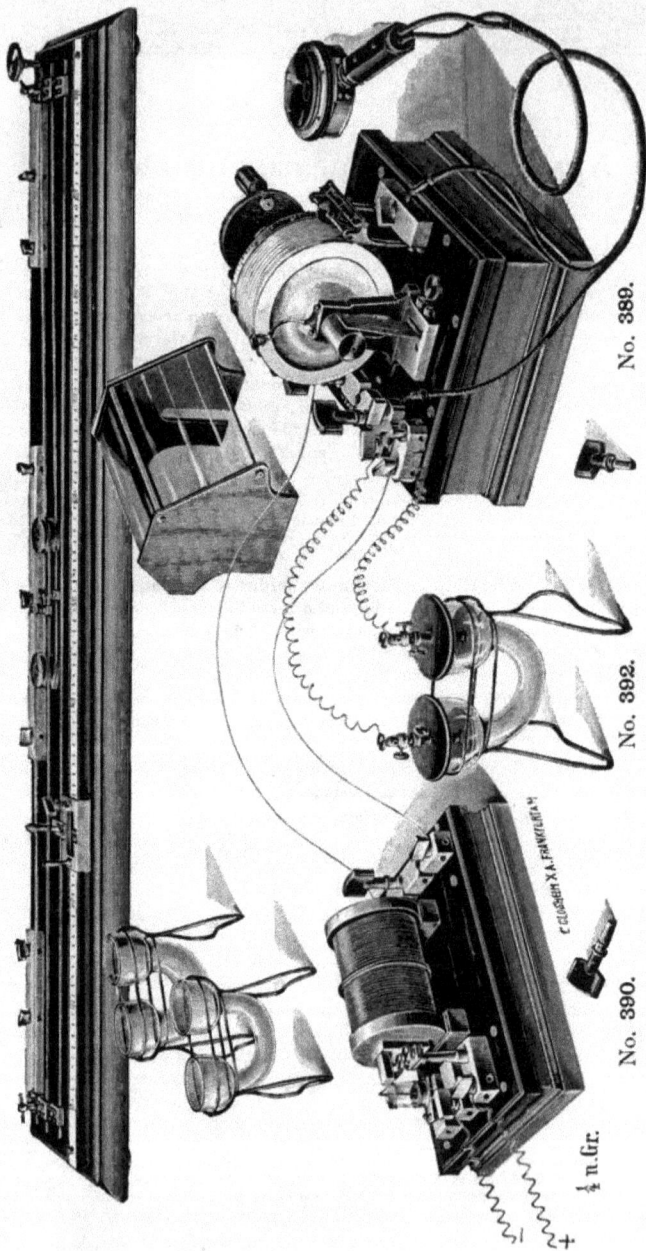

No. 389a. Bridge with stretched wire.

No. 389.

No. 392.

No. 390.

½ n. Gr.

Kohlrausch Apparatus to determine the resistance of electrolytes.

Standard Resistances.

All our **resistance boxes with plug contacts** are made with the greatest care and only the best materials are used and they are always repeatedly calibrated. The wire employed is an alloy with a practically negligeable temperature coefficient and only used some considerable time after it is coiled. The coils are double*) wound and to obviate any faults which might occur from the hygroscopic nature of the silk covering, each coil is kept for a long time in a hot air bath and then, when all damp is certainly removed, varnished, or else soaked hot in melted paraffin which penetrates right through the coil.

The inner and outer wires of two neighbouring coils are not connected to one common lead but are each fitted to their own lead which then terminates separately in the metal block which may be considered as without resistance. It is consequently evident that the sum of each separately measured resistance equals the total combined resistance.

Each metal block is immoveably mounted on the polished ebonite plate by screws and pins and has a conical hole in the side to which a terminal headed plug is fitted, any one of the resistances therefore can be used whilst obviating the joint resistance of the contact plugs.

These latter are made very heavy to reduce their resistance in circuit and are sharply coned to prevent their setting fast under variations of temperature, their ebonite tops are of convenient shape, solidly moulded on and absolutely immoveable, care being taken, that thermo-currents from handling will not occur. The cases are made with large holes, which can be closed if required, to enable the in- and external temperature to compensate rapidly and for the purpose of introducing a thermometer. The coils are calibrated to a fractional part of one onethousandth, and the official Certificate from the **Imperial German Laboratory** will be supplied, if desired for any of these standards, at cost price.

---

*) The higher resistances will willingly be supplied, wound on Chaperon's inductionless system, through which the electrostatic capacity of the resistances is considerably decreased and the resistances better adapted for telephonic measurements. In this case the prices will be slightly increased.

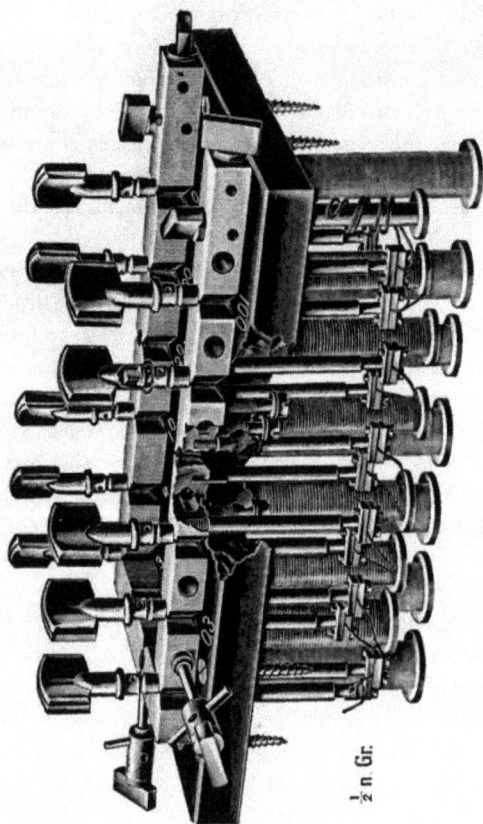

Standard resistances.
(Showing details of mounting.)

$\frac{1}{2}$ n. Gr.

No. 394a.

Standard Resistance Boxes. Series arrangements.

No. 398.

No. 394.

No. 402.

¼ n.Gr.

Standard Resistance Boxes: Series arrangements.

The consecutive arrangement*) of resistances adopted in these series rheostats of 1, 1, 2, 3, 4, 10, 20, 30 et seq. enables each resistance to be accurately calibrated and checked by the comparison of a higher with the combined sum of two smaller resistances, for which purposes the jack plugs which are supplied with the box will be found very useful. The additional resistance of $^1/_{10}$ of the smallest resistance in the series Rheostat is intended for interpolation purposes. All the resistances are marked on both sides of the blocks and can therefore be easily read from either side.

Standard Resistance Boxes with 6 plugs reading up to.

No. 393.	I	II	III	IV	V	VI	VII
Total combined resistance	1.11	11.1	111	1110	11100	111000	1.11 megohm
Price M.	90	80	80	90	120	200	450

Series Rheostats commencing with resistances of 0.01, 0.1, 0.1, 0.2, 0.3, 0.4, 1, 2, 3 et seq.

No.	394	396	398	400	402
Number of plugs	14	16	18	20	22
Highest single resistance ohms	40	200	400	2000	4000
Total combined resistance ohms	111.11	411.11	1111.11	4011.11	11111.11
Price M.	160	180	210	240	270

Series Rheostats commencing with 0.1, 1, 2, 3, 4, 10 et seq.

No.	394a	396a	398a	400a	402a
Number of plugs	10	12	14	16	18
Highest single resistance ohms	40	200	400	2000	4000
Total combined resistance ohms	111.1	411.1	1111.1	4111.1	11111.1
Price M.	130	150	180	210	240

As those figured in small type are seldom asked for, they are not kept in stock, and will therefore require a longer time for delivery.

---

*) Any other combination can be supplied, and the resistances adjusted in Siemens units instead of ohms, if desired.

Standard Resistance Boxes
with decades of equal or varying resistances.
Branch Resistance Boxes.

¼ n Gr.

No. 403c.   No. 393.   No. 403a.

Decade Rheostats of 10 equal resistances, the circuit being made by a single plug; a second plug is employed to vary the resistance, the circuit remaining unbroken, and thereby eliminating errors due to variations of resistance in the numerous plug contacts of the Series arrangement.

No. 393a	I	II	III	IV	V	VI	VII
Single resistances . ohms	10×0.1	10×1	10×10	10×100	10×1000	10×10000	10×100000
Price M.	160	150	150	160	180	250	500

Decade Rheostats with resistance values of 2×1 and 4×2, thus enabling all values from 1—10 to be combined by the use of two plugs; they are consequently more convenient.

No. 393b	I	II	III	IV	V	VI	VII
Total Resistance . ohms	1	10	100	1000	10000	100000	1 megohm
Price M.	130	120	120	130	150	210	460

Branch Rheostats for use with bridges. The two arms of these rheostats are connected by a plug; by means of the two additional branch plugs they can be used independantly as branches.

No. **403. Branch-Rheostat** with pairs of 1, 10, 100   ohms   Price M. **110.—**
  „ **403a.**   „   „   „   „ 10, 100, 1000   „   „   „ **115.—**
  „ **403b.**   „   „   „   „ 1, 10, 100, 1000   „   „   „ **130.—**
  „ **403c.**   „   „   „   „ 1, 10, 100, 1000, 10000 „ „   „ **200.—**

Shunt and Additional Rheostats for use with galvanometers.

$\frac{1}{4}$ n.Gr.

| No. **404b.** | No. **408a.** | No. **404.** |

Shunt Rheostats to reduce the sensitiveness of galvanometers, wound with copper wire resistances and inductionless winding; with resistances $1/9$, $1/99$, $1/999$, and $1/9999$ of the resistance of the instrument for which they are intended; assuming that the smallest resistance of the shunt is not much below 0.1 ohm. With a short circuiting plug.

No. **404.** Shunt Rheostat with 2 resistances ($1/9$ and $1/99$) Price M. **120.—**
  „ **404a.**   „   „   „   3   „   (ditto + $1/999$)   „   „   **135.—**
  „ **404b.**   „   „   „   4   „   (ditto + $1/9999$)   „   „   **160.—**

Additional Rheostats with from 2—4 resistances, wound with Constantan, of 9, 99, 999, 9999 times the resistance of the instrument for which they are intended and of which it is desired to extend the range; assuming that the highest resistance does not exceed 100000 ohms. With a short circuiting plug.

No. **408.** Additional Rheostat with 2 resistances (9 and 99) Price M. **120.—**
  „ **408a.**   „   „   „   3   „   (ditto + 999)   „   „   **140.—**
  „ **408b.**   „   „   „   4   „   (ditto + 9999)   „   „   **210.—**

Standard Rheostat for large currents.

393c I. Rheostat with 10 resistances, each 0.1 ohm, in Constantan, with large copper and heavy nickelled terminals for use with mercury contacts; suitable bow and comb shaped connectors are supplied to couple the resistances either in series or parallel; in the latter case a current strength of 100 ampères is admissible. The box can be used as a petroleum bath.   Price M. **500.—**

Resistances for ordinary testing,

with plug contacts.

1/5 n. Gr.

No. **393d.**   No. **398d.**

Resistances, in which the accuracy ot calibration has not been carried so far as in those previously described, are for a great many purposes however, sufficiently accurate, and we therefore keep a stock of both single resistances and sets of four mounted on ebonite; two, three, or four, of these sets can be mounted together in one box, and each set can be connected together by loose contact strips. The various resistances are introduced into the circuit by withdrawing the respective plugs, and each metal block is bored out to take a screw clamp, so that any shunt circuit can be arranged as previously described on page 38. These resistances are calibrated to within $^1/_2$ per cent.

Single resistances as above in wood case with short-circuiting plug.

No. **409d.**	I	II	III	IV	V	VI	VII
Ohm	0.1	1	10	100	1000	10000	100000
Price . . M.	**20**	**18**	**18**	**18**	**20**	**30**	**50**

Rheostats as above, series arrangement

No. **393d** II with resistances of   1,   2,   3,   4 $=$ 10 ohms, Pr. M. **36**
„   „   III   „   „   „ 10,   20,   30,   40 $=$ 100   „   „   „ **36**
„   „   IV   „   „   „ 100, 200, 300, 400 $=$ 1000   „   „   „ **40**
„   „   V   „   „   „ 1000, 2000, 3000, 4000 $=$ 10000   „   „   „ **40**
„   „   VI   „   „   „ (10, 20, 30, 40) $\times$ 1000 $=$ 100000   „   „   „ **70**

Sets of the above combined in one case
No. **394d** with Nos. 393d II and III,   total   110 ohms,   Price M.   **70**
„   **398d**   „   „   393d II, III and IV,   „   1110   „   „   „   **100**
„   **402d**   „   „   393d II, III, IV and V,   „   11110   „   „   „   **130**

Rheostats for branch circuits, of similar design as above but calibrated more accurately than the series rheostats
No. **403d** I with 4 resistances   1, 1, 10, 100 ohms,   Price M. **50**
„   „   II   „   4   „   10, 10, 100, 1000   „   „   „ **50**

Bridges made from these resistances are described on page 47.

Decade Rheostats with switch contacts.

¹/₄ n. Gr.

No. 402a.

The **Decade Rheostats** with switch contacts are each made up of 10 equal resistances of 1, 10, 100, or 1000 ohms, calibrated within ¹/₂ per cent. As compared with series rheostats they have the advantage that the resistances can be varied equally in parts of the unit of resistance in each decade; whilst in the series arrangement the variations can not be effected in this manner.

Decade Rheostats:

No. **393e** II with resistances 10×1   total   10 ohms, Price M. **55.—**
 „   „   III   „   „   10×10   „   100   „   „   „ **55.—**
 „   „   IV   „   „   10×100   „   1000   „   „   „ **60.—**
 „   „   V   „   „   10×1000   „   10000   „   „   „ **60.—**
 „   „   VI   „   „   10×10000 „   100000   „   „   „ **90.—**

Sets of above combined in one case.

No. **394e** with decades 393e II and III, total 110 ohms, Price M. **105.—**
 „ **398e**   „   „   „   II, III a. IV,   „   1110   „   „   „ **150.—**
 „ **402e**   „   „   „   II,III,IV,V, „   11110   „   „   „ **200.—**

Branch Resistances with switch contacts each with 2×3 resistances, for various arrangements of circuit, accurately adjusted.

No. **403e** I, with 2 switch keys, each for 1, 10, 100 ohms, Price M. **80.—**
 „   „   II, „ 2   „   „   „   „   10,100,1000   „   „   „ **80.—**
 „   „   III, „ one key for 1, 10, 100 and the other for
        10, 100, 1000 ohms . . . .   „   „ **80.—**

Bridges made from these decade rheostats with switch contacts are described on page 47.

Wheatstone Bridges with Standard resistances,
with plug contacts.

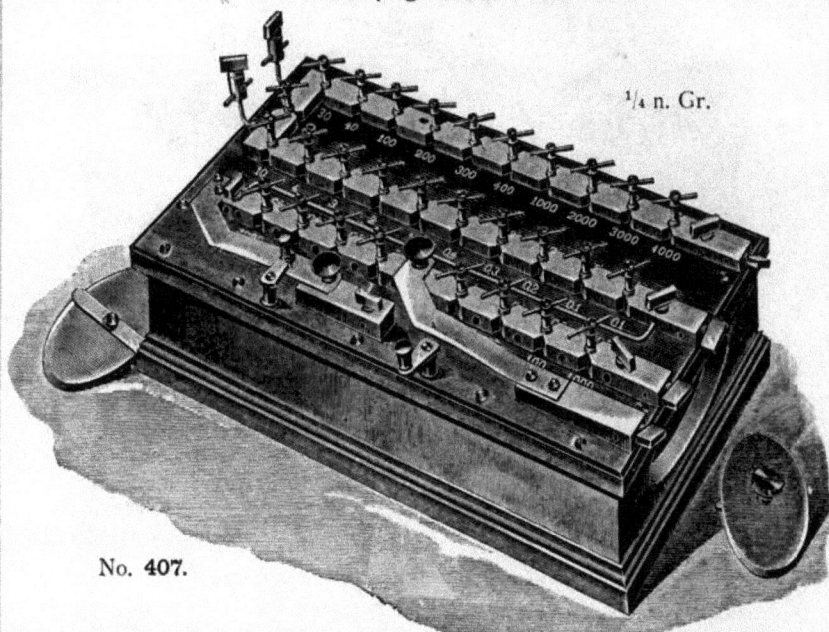

¼ n. Gr.

No. **407.**

These **bridges** are made up with the Standard rheostats Nos. 398 or 402 (page 41) and the branch reostats Nos. 403 or 403b combined with the requisite battery and galvanometer keys. The various parts are mounted together on one piece of ebonite and are so arranged that the terminals for the unknown resistance are brought as close together as possible. In order to check the resistances used in the branches, the two arms can be exchanged one against the other, in which case the long connector between the series and branch rheostats is removed and a short one employed at right angles to the former.

No. **405. Standard Bridge;** with 23 plugs,*) resistances of 0.1, 0.1, 0.2, 0.3, 0.4. 1, 2, 3, 4, 10, 20 &c. up to 400, total 11 111 ohms, besides the pairs of 1, 10, 100 ohms; suitable for measurement up to, approx. 110 000 ohms.   Price M. **330.**—

No. **407. Standard Bridge;** with 29 plugs*) resistances as above but up to 4000, a total resistance therefore of 111 111 ohms, besides the pairs of 1, 10, 100, 1000 ohms; suitable for measurements up to, approx. 11 million ohms.   Price M. **420.**—

If sufficient time for manufacture is accorded, we can supply bridges with any desired combination of resistances.

*) The plugs have ebonite handles (as described and illustrated on pages 38 and 39) and not capstan heads as shown.

Bridges for ordinary testing
with plug and switch contacts.

No. 407e.

¼ n.Gr.

No. **405 d. Ordinary Bridge with plug contacts,** made up with the Series rheostat No. 398 d with 3 rows of together 1110 ohms and the branch rheostats Nos. 403 d I or II (page 44) all mounted in one case — without keys — suitable for measurements up to, approx. 100000 ohms.
Price M. **150.**—

No. **407 d. As above,** but with No. 402 d and 4 rows of resistances, together 11 110 ohms and the branch resistances No. 403 d I or II (page 44) suitable for measurements up to approx. 1 million ohms. Price M. **180.**—

No. **405 e. Ordinary Bridge with switch contacts:** a rheostat No. 398 e with 3 decades, together 1110 ohms, and the branch resistances Nos. 403 e I, II, or III (page 45), and double-key contact; all mounted together in one case, and suitable for measurements up to 100000 or 1 million ohms.*)
Price M. **250.**—

407 e. As above, but with rheostat No. 402 e with 4 decades, together 11 110 ohms, and the branch resistances Nos. 403 e I, II or III (page 45) and double-key contact, suitable for measurements up to 1 or 10 million ohms.
Price M. **300.**—

Simple Wheatstone Bridges, reading the resistance direct on measuring wire. See pages 60 and 61.

---

*) The use of a sufficiently sensitive galvanometer (reflecting galvanometer) is implied if these high ranges are required.

No. 506.
Portable Universal resistance-measuring apparatus.

ca. $\frac{1}{8}$ n. Gr.

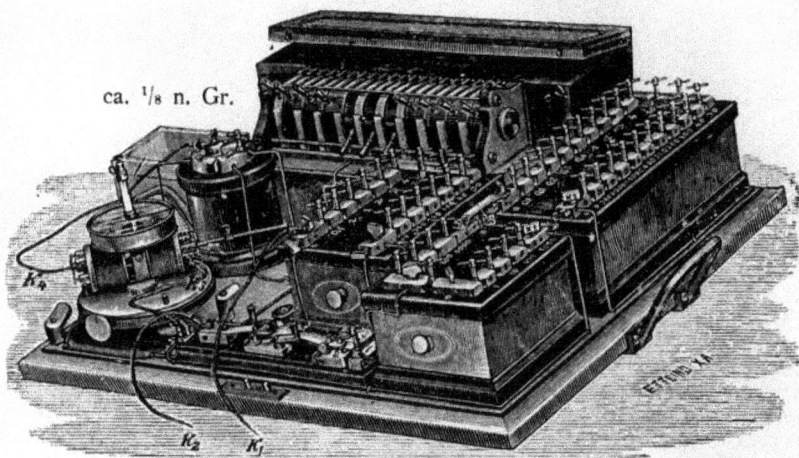

No. **506**. This **Set of measuring instruments** for use on every range of resistance is made up of the series rheostat No. 402 combined resistance 11111.11 ohms, the branch rheostat No. 403 b with pairs of 1, 10, 100, 1000 ohms; the standard resistance No. 553 of 0.01 *); a differential galvanometer No. 367 with one pair of coils of 100 ohms and a second pair of 4000 ohms and a shunt resistance No. 404a with 3 reduction resistances to suit the latter coils. A double key and plug switch for the various measuring arrangements; a battery commutator switch No. 496 to combine the dry cell battery (10 cells) in series, parallel, or separately, as required; and the battery above mentioned in separate case. With the exception of the battery, all the above apparatus is mounted and connected on an oak tray which is fitted with strong handles for carrying purposes, and with a lock-down cover to protect the various instruments. The combined apparatus is extremely well suited for accurate testing on widely extended systems of mains from central supply stations and is arranged for:

1. Measuring resistances by means of the Wheatstone bridge from 0.1—1 million ohms.
2. Measuring low resistances by Kirchhoff's shunt method and differential galvanometer; range from 1 down to 0.0001 ohm.
3. Measuring high resistances, such as cable insulation, up to 10 Million ohms by the direct deflection method.

The various circuit combinations for each of these arrangements are fitted under glass on the tray. Complete working instructions are supplied with each apparatus.

Price M. 1600.—

If desired we shall be pleased to modify this combination in any way; especially as regards the galvanometer No. 367 for which No. 367a or No. 535 can be substituted, thereby extending the range for insulation testing in a corresponding degree. The difference in value between either of these galvanometers and No. 367 has to be, in this case, added to the above price.

---

*) In lieu of the second series rheostat shown on the engraving.

No. 641. Portable apparatus to measure resistances and potentials.

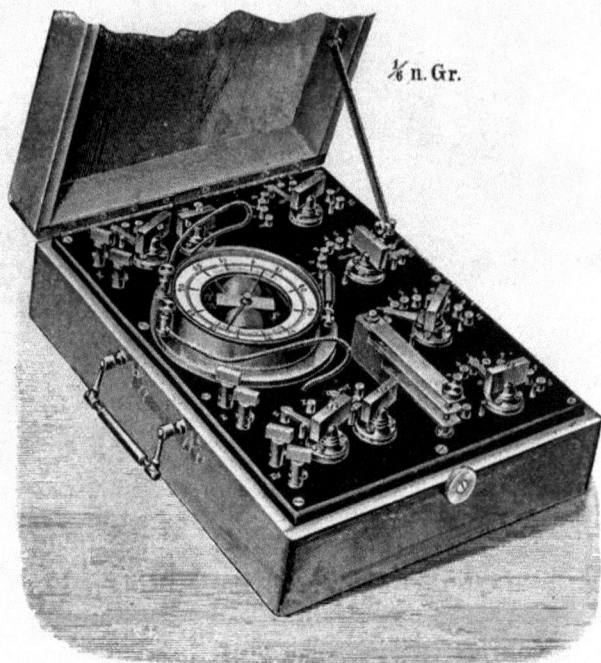

⅙ n. Gr.

No. 641. Portable apparatus to measure resistances and potentials especially suitable for testing cells. The oak case contains a Wheatstone bridge with switch contacts similarly arranged to No. 407e and a dead-beat galvanometer similar to either No. 366 or No. 536a and arranged to turn in its seat, together with commutator to reverse the battery poles and switch for the two different kinds of measurement. The whole apparatus is mounted ready for use on the ebonite base. Range, when used to measure resistances, from about 0.1 to 1 Million ohms. Price M. **580.**—

Instructions or use. **Resistance measurements.** The unknown resistance is connected to terminals **X**, the battery to the terminals **E**, the switch **U** on the contact **E**. **Ex** remain open. The switch contacts of the comparative resistances **a** and **b** are adjusted to a suitable proportion, and the decade resistances **R** are then varied until the galvanometer gives no deflection when the double-key is pressed down. Then $x = \dfrac{b}{a} R$.

To measure **potentials** (substitution method) a standard cell is connected to **E**, the unknown potential to **Ex** and the keys **a** and **b** on to the contacts **Ex**. **X** is left open. The switch **U** is then brought on contact **E**, and the contacts of the decade resistances adjusted until the galvanometer gives a suitable deflection when the double-key is pressed. **U** is then turned on to **Ex** and the decade resistances again adjusted until the same deflection as before is attained. The two potentials are then in inverse proportion to the two resistances determined.

No. 508.
Low resistance bridge for specific resistance measurements.

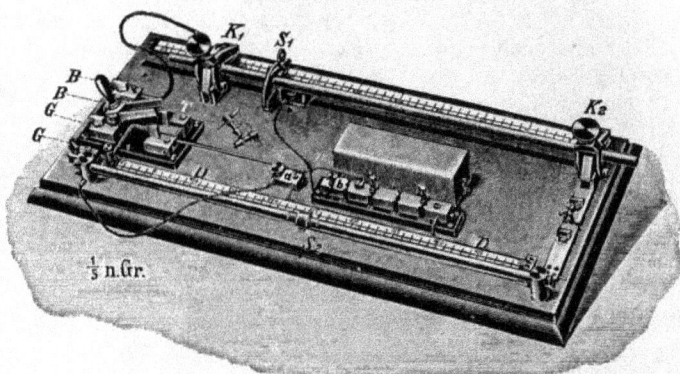

⅓ n.Gr.

No. 508. This low resistance bridge is specially adapted for measuring the relative conductivity of various materials and for testing arc-light carbons &c.; it is used on the Wheatstone bridge principle, but the resistances of the contacts are eliminated by double readings. Two universal clamps, taking rods or wires of any section, grip a certain definite length of the material (up to 530 mm) and the reading is taken on the calibrated wire and requires but little calculation. A double key is used to close the battery and galvanometer circuits in succession. Range from about 0.00001 to 7 ohms. Price M. **300.—**

Directions for use. A straight piece of the material to be tested is inserted between the clamps K_1 and K_2; a sensitive galvanometer (such as No. 367a, 535, or 535:) is coupled between GG, and any convenient source of current, preferably accumulators, to BB. — Two of the branch plugs for instance 10:10, 10:100, 10:1000 according to ratio required, are withdrawn and the sliding contact S_1 is brought near K_1 and S_2 adjusted until opening switch T does not alter the galvanometer reading; S_1 is then brought near K_2 and S_2 again adjusted until no deflection occurs. The unknown resistance of the piece of material included between the two positions of S_1 is equal to the ratio of the two branch resistances in circuit multiplied by the resistance of that part of the calibrated wire included between the two positions of S_2 which is read off by subtracting the two resistances then found. The length of X, or in other words the distance between the two positions of S_1 is read off exactly on the second scale which is divided into millimetres. If it is desired to measure other low resistances such as, for instance, the resistance of a coil of insulated wire, or the windings on a dynamo, the two ends of the unknown resistance are connected as before with K_1 and K_2, using for this purpose if necessary, connecting wires of suitable size. Instead of the sliding contact S_1 being used a flexible wire is now connected to b and contact made thereby with K_1 (or the end of the unknown resistance connected therewith) and the position of the sliding contact S_2 is then determined and read off; the flexible is then transferred to K_2 and S_2 again adjusted until the galvanometer does not deflect when the circuit is repeatedly made and broken.

The current strength admissible depends on the section of the material under examination. The instrument will stand a maximum current of about 5 ampères for a short time, and it is advisable to check this by means of an ammeter such as No. 380a II.

No. **508a.** **Thomson's** (Lord Kelvin's) **Double-Bridge.** By the addition of a second comparative rheostat and another moveable clip contact to define accurately the length of the piece of material under examination, the low resistance bridge No. 508 can be used without further alteration as a Thomson's double-bridge. It has the same range as No. 508 and is especially suitable for measuring irregularly formed resistances, such as turns on armatures, &c. Price M. **375.—**

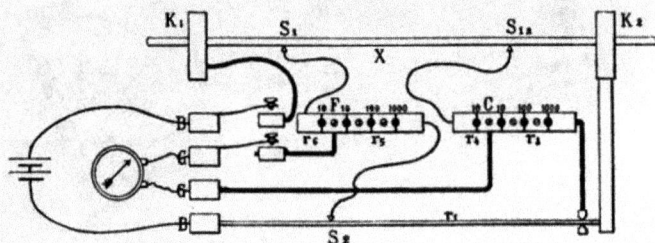

Instructions for use. The unknown resistance is coupled to K_1 and K_2, and a very sensitive galvanometer to GG; the source of current to BB. By withdrawing plugs in the comparative resistances any suitable ratio can be selected, the contacts C and F being adjustable to any desired position between the various resistances for this purpose. The moveable contact S_2 is then adjusted along the stretched measuring wire, which is graduated into thousandths of an ohm, until the successive contact key may be repeatedly pressed down without causing a deflection of the galvanometer. Then, taking the ratio between the resistances $\dfrac{r_4}{r_3} = \dfrac{r_6}{r_5}$ as n, the resistance x in circuit between S_1 and $S_{1a} = r_1 \cdot n$.

For measuring very low resistances by Kirchhoff's differential shunt-method the following apparatus are necessary. A differential galvanometer such as No. 367 or 367a pages 8 and 9, No. 535 or 535a pages 14 and 15, or instead of one of these either No. 371 or 371a pages 10 and 12; a standard branch resistance such as No. 553 or 554 page 30; a standard rheostat with tenths of an ohm such as No. 398 or 402 page 41; and a battery key such as No. 492 or commutator No. 493a page 63.

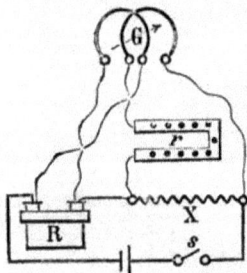

The unknown resistance X is introduced into the circuit in series with the standard resistance R and the terminals for these two resistances are also connected each to one half of the differential galvanometer. The rheostat r is coupled in whichever of these two circuits is of the lowest resistance, and additional resistance thereby inserted, until the galvanometer does not deflect when the circuit is made. The two low resistances in the main circuit are then in the same ratio as the two larger resistances in the shunt circuits.

It is more convenient to use a rheostat in each of the galvanometer circuits, but in this case the second rheostat need not be arranged with tenths of an ohm. No. 398a, page 41, is quite sufficient for the purpose.

No. 614.
Ohmmeter to read direct in Ohms.

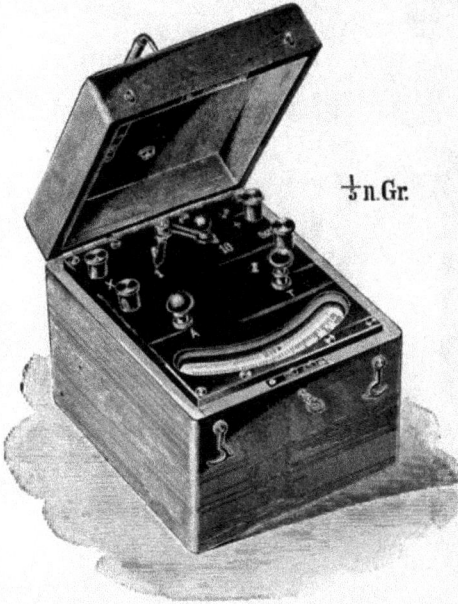

$\frac{1}{3}$n.Gr.

No. **614. This Ohmmeter reads direct in Ohms** independantly of the battery strength employed. It consists of a specially designed differential galvanometer, which is unaffected by the earths magnetism, though without very sensitive astatization, and which indicates by a pointer and scale the resistance of the circuit in ohms without any further calculation. The percentage of error is approximately the same throughout the whole range on the scale, and if used with a potential of 50 to 100 volts*), for example, the potential at which an installation is supplied, is about 1%. It has a range from one to 100 fold and by means of a shunt, this range can be extended to 1000 fold. It is designed especially for testing the insulation of an electric light installation with a range accordingly from 1000 to 1 million ohms. The terminals for the unknown resistance are to the left of the instrument and battery terminals on the right; they are cased in ebonite, and, together with circuit key and automatic stop for travelling purposes, are mounted on polished ebonite base, which is enclosed in an oak case, the lid being fitted with handle to carry the instrument.

Price M. **250.**—

This instrument can be supplied with any desired smaller range and consequently more open scale; in which case the same degree of accuracy is attainable with smaller battery power.

*) **60 small Cells** No. 501 c (page 66) in oak case of same size and shape as for Ohmmeter . . . . Price M. **120.**—
or a **Magneto** for continuous current in a similar oak case. „ „ **50.**—

No. 509. Portable apparatus for measuring high resistances (Testing cable insulation &c.)

$\frac{1}{8}$ nat. Gr.

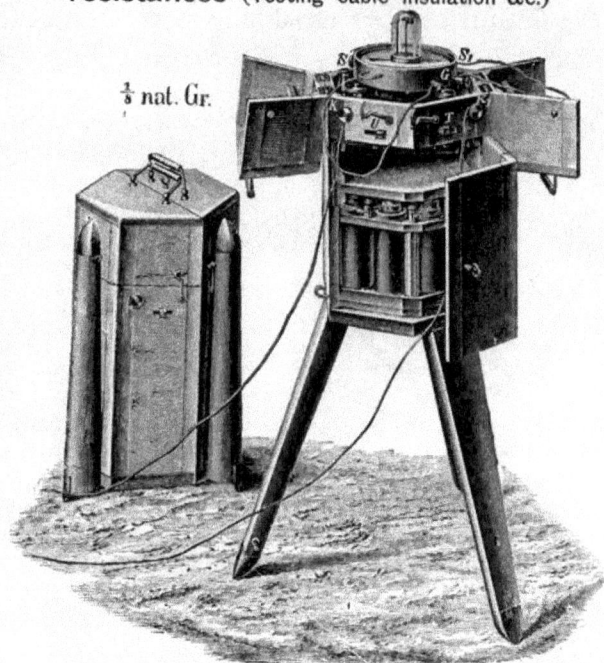

No. 509. **Portable apparatus for measuring insulation resistance** comprising an accurately calibrated dead-beat galvanometer of high resistance and shunt for $^1/_{10}$ sensitiveness, a dry cell battery of approximately 70 volts, a key to put the galvanometer in circuit either with or without shunt, and a two way switch to put either a comparative resistance of 100000 ohms or else the unknown resistance in circuit with the galvanometer. This apparatus is all combined in a very strong lock up case for travelling purposes, fitted with folding metal shod tripod legs, as shown in cut, and weighs approximately 14 kilogramms. It is especially suitable for use in cable laying. Whilst using the method of direct deflection, the readings are not affected by variations of battery potential. Range from about 1000 to 15 million ohms. Price M. **375.**—

Instructions for use. The unknown resistance is connected to terminals K and K_1 and the two way switch is placed on the contact marked VERGL. (comparative resistance) and the key T pressed and moved on to the contact marked. GALV.; a deflection is obtained of say 35°. The two way switch U is then turned on to the contact marked UNBEK. (unknown resistance) and the key again pressed down, but only moved on to the contact marked GALV. if the galvanometer needle is but slightly deflected, that is, if no short circuit is present; a deflection is then obtained of say 8.5°. The values corresponding to these figures have then to be determined from the curve given with each instrument, the readings being ordinates on the curve whilst the values are shown on the abscissa (reciprocal of the resistances) they are 120 and

24.5. Then the unknown resistance $x = \dfrac{120 \times 100000}{24.5} = 489796$ ohms, from which result, in extremely accurate work,

the resistance of the galvanometer (1000 ohms) would have to be subtracted.

No. **509a.** **A similar apparatus as above** but arranged with **Ohmmeter No. 614** reading to 1 megohm, thus eliminating all further measurements or calculations. Price M. **420.**—

No. 509 b.

Portable apparatus for measuring high resistances (large size).

No. 509 b. **Portable apparatus for measuring insulation resistances** (large size). A weather proof case in oak, tongued and grooved, of similar shape but larger than No. 509 is so arranged that when the tripod legs are folded up it can, by means of shoulder straps, be carried about and erected where required by one man. In this case is a larger sized dry cell battery of over 100 volts and, in lieu of the needle galvanometer, the astatic reflecting galvanometer with telescope No. 367a (see page 9) with the telescope bracket arranged to turn over for conveniently housing the instrument within the case. A switch which at the same time serves as a selector for 25, 50, 75, or 100 cells, a multiple contact switch for the 3 galvanometer shunts, and a similar one for changing the comparative resistance of 100000 ohms with the unknown resistance, and the requisite terminals for the latter, are all mounted on a polished ebonite plate. This combination for measuring resistances by the direct deflection method affords a very high range — maximum 500 megohms — is also applicable to the measurement of comparatively low resistances, and is therefore a very good substitute for a travelling testing-cart when laying cables. Price M. 800.—

Instructions for use: When the mirror galvanometer has been set up and adjusted on the open lid of the apparatus so that the magnet swings freely and the plane of the coils lies approximately in the magnetic meridian (operations which can be performed by quite unskilled hands) the unknown resistance is connected to the terminals K and K_1 and the switch U is turned on to contact „Unbek." (unknown resistance), the contact key T which serves also as battery-selector being turned until a suitable deflection is obtained when it is depressed. It is advisable, in cases where even the approximate resistance is unknown, to work with the least sensitive arrangement of the galvanometer, switching in by means of U_1 the smallest shunt, as the sensitiveness can always be increased by afterwards turning the switch U_1 to the left if, in its first position, the galvanometer is not sufficiently sensitive to give a suitable deflection. U is then turned on to „Vergl." thereby placing the comparative resistance of 100000 ohms in circuit in lieu of X, adjusting the sensitiveness of the galvanometer by means of U_1 until its deflection is not too different from that previously obtained, when depressing the contact key T, which of course must be in the same position as in the previous measurement. These two readings then give

$$x = \frac{100000 \times a_v \; s_x}{a_x \times s_v}$$

a_x and a_v being the deflections obtained with the comparative and unknown resistances respectively and s_v and s_x the respective degrees of sensitiveness as read off at U_1.

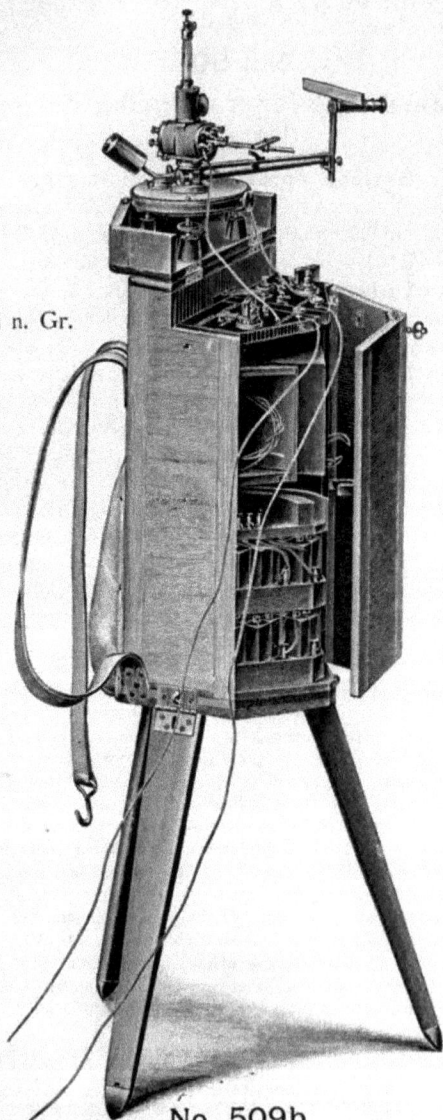

ca. $^1/_8$ n. Gr.

No. 509b.
Portable apparatus for measuring high resistances (large size).

No. 507.
Cable testing cart and tent.

A well built two wheeled hand truck serves to carry all the various instruments, some of which are removeable whilst others are mounted inside the truck and are accessible on raising the lid and dropping the flap at back of truck. The truck is fitted with an easily erected tent to protect all the instruments when in use, and also with adjustable iron props which enable it to be firmly fixed wherever required; provision is also made for carrying the light tripod stand on which the mirror galvanometer and telescope (No. 367a) is fixed, to be used under the tent whenever measurements are being made. The apparatus contained in the truck comprises a bridge with switch contacts, and an instrument for insulation measurements by the direct deflection method, all mounted with the requisite keys, commutators, battery selectors &c., but this arrangement can be modified as desired. The various instruments are mounted on a sliding tray, so that if not in use, they can be pushed back from the hinged back of the cart, thus affording space for other apparatus, more especially a standard voltmeter and ampèremeter. A small battery for use with the bridge and a larger one of over 100 volts potential for insulation testing are arranged in the less accessible parts of the truck, and also a magneto with bell and galvanometer for testing house wiring; a lock up box is also fitted in front between the shafts to take all requisite tools.

Price (according to the outfit) from . . . . . . M. **2500.—**

No. 507. Cable testing cart and tent.

VI. Apparatus for testing lightning conductors and earth connections for telegraph and telephone stations.

No. 450. Nippoldt's telephone bridge.

$^1/_2$ n. Gr.

No. **450. Nippoldt's telephone bridge** for testing the resistance to earth of lightning conductors. In this instrument a bridge is combined with a telephone in a clock shaped case so that the calibrated wire is thereby protected. The sliding contact is fixed to the graduated dial which can be turned round as required and the resistance read off direct in ohms. A comparative resistance and commutator are also fitted in the case, the latter is connected to a 5 cored flexible cable in which the two black and green covered leads are connected to the resistance to be measured, and the green to the source of current, whilst the brown, if necessary, is coupled to an earth contact. The range is from 0.1 to 100 ohms; resistances between 100 and 200 ohms can also be estimated. Price M. 80.—

The requisite current for testing lightning conductors and measuring the resistance of electrolytes is best obtained from a small induction coil with high speed make and break (high note) fed from 1 or 2 cells.

Terminals G and G₁ are provided on the side of the case if a galvanometer is to be connected for determining the resistance of wires or other materials within the limits above mentioned; in this case the short-circuiting strip h is opened to cut the telephone out of circuit.

$^1/_2$ n. Gr.

No. **451. Simple pocket galvanometer** with index pointer. This instrument is especially suitable for use with the lightning conductor set No. 452 as its general arrangement renders it extremely portable; it is at the same time useful for many other purposes. A small thimble magnet having sufficient play within the copper damper to obviate the necessity for levelling screws is suspended on a short raw silk fibre, whilst a milled edged nut on the side of the case enables the strain to be taken off the fibre for travelling purposes. Price M. 50.—

Sensitiveness: 1° deflection = 0.0001 ampère.

This instrument can, in combination with the telephone bridge above described, be used for determining the resistance of wires and other solid materials, and especially in those cases where induction occurs, such as coils and the like. The terminals of the two green cords of the telephone bridge No. 450 are connected to those marked P on apparatus No. 452, or, in the event of this not being available, to those of a Leclanché or dry cell, or cells in series. In this case the lever on the side of the telephone bridge is placed on the mark I and the brown cord is not used.

No. 452.
Complete lightning conductor testing set.

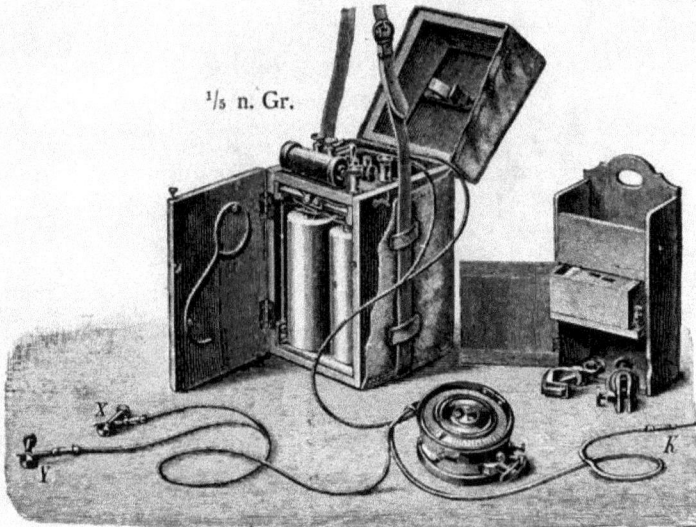

1/5 n. Gr.

No. **452. Complete lightning conductor testing set,** comprising Nippoldt's telephone bridge No. 450, with induction coil for generating alternating currents, and one to two dry cells to work same, with circuit switch; fitted in walnut case, with drawer for connectors, galvanometer (No. 451) &c., with leather case and shoulder straps. Price, exclusive of galvanometer and connectors . . . . M. 120.—

Directions for use. The unequal ends of the black and green flexible twin wires are connected to the conductor the resistance of which is to be measured, and the ends of the green flexibles to the terminals marked S on the induction coil. The switch is then moved on to the contact marked TEL. when a humming sound will be heard in the telephone which either increases or decreases by turning the dial, until the humming either entirely or almost entirely disappears. The resistance is then read off on the dial. The lever at side must be on I and the brown flexible cord is not used.

If the galvanometer is to be used to determine the resistance, the ends of the green flexibles are connected to the terminals marked P and the switch is moved to the contact marked GALV.; as for the rest see page 58.

Full instructions for determining the resistances of joints and to earth of lightning conductors are supplied with each set.

Accessories:

1. **Folding earth plate,** made up of 10 tin sheets each 25×25 cm connected by hinges, and therefore extremely portable, with terminal . . . . Price M. 15.—
2. **Earth contact,** a steel drill with handle, a substitute for a second earth when testing lightning conductors which have only one earth. Price M. 6.—
3. **Connectors** for coupling on to stranded lightning conductors or to strip do. (see illustration above) . . . . Price each M. 1.75
4. **Couplings** for stranded lightning conductors . . Price M. 2.25

In certain cases the earth lead of a ligthning conductor must be separated from that part above ground, and, to avoid cutting the conductor which would necessitate making and soldering a new joint, the use of the coupling illustrated at side is to be recommended.

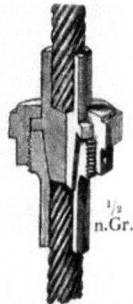

1/2 n.Gr.

No. 391. Kohlrausch's universal bridge.

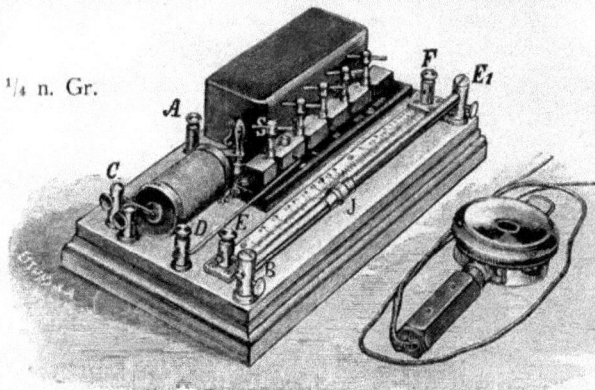

$^1/_4$ n. Gr.

No. **391.** **Kohlrausch's universal bridge** differs from most other arrangements as the readings are given on a scale direct in ohms and no reference to tables is necessary. The rheostat has comparative coils of 1, 10, 100 and 1000 ohms.*) When used with a suitable galvanometer such as No. 366, page 7 or No. 536a, page 15 this instrument will measure, with sufficient accuracy for all industrial work, resistances of solid conductors (such as wires) from 0.1 to 10000 ohms.

If, instead of the galvanometer, a suitable telephone and alternating current be employed, for which purpose a small induction coil is mounted with the apparatus, resistances of electrolytes, for example the internal resistances of cells, and the resistance to earth of earth plates of lightning conductors can also be determined. Price M. **130.**—

Directions for use: Measuring resistances of solid conductors. The unknown resistance is connected to the terminals D and E; and the battery (2—3 cells) to A and B; the galvanometer to E₁ and F. The plug S is withdrawn and one of the plugs 1, 10, 100, 1000. The switch a is moved on to the contact marked GALV. and the pointer J is adjusted until the galvanometer does not deflect when circuit is made and broken. The figure shown by pointer J has then to be multiplied by the value of the comparative resistance in circuit, which should always be selected to bring the pointer as near as possible in the middle of the scale.

Measuring the resistance of electrolytes. The battery for working the induction coil is connected to A and C, and a telephone instead of the galvanometer to E₁ and F, and plug S inserted. The switch a is moved on to contact marked TEL. and the pointer adjusted until the telephone is silent. The resistance is read off and multiplied as above described.

No. **1.** **Watch shape Telephone** for determining the resistance of electrolytes, for use with universal bridge No. 391, with the fields specially wound for the above purpose. Price M. **15.**—

No. **1a.** **As above,** with ebony handle. „ „ **20.**—

No. **2 & 2a.** **Telephones,** differentially wound. Price M. **20.**— or **25.**—

*) If **specially ordered,** the Kohlrausch bridge No. 391 and 391a will be supplied with rheostats of 0.1, 1, 10, 100 ohms (instead of 1, 10, 100, 1000) thus rendering them especially suitable for low resistance measurements.

No. 391a. Simple Kohlrausch bridge.

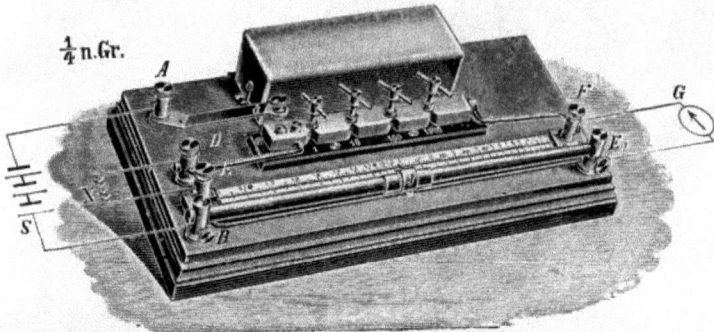

¼ n.Gr.

No. 391a. **The simplified Kohlrausch bridge** is the same as the Universal bridge No. 391 but without the induction coil, it is therefore only suitable for direct reading resistance measurements of solid conductors within a range of 0.1 to 10000 ohms, (if a sufficiently sensitive galvanometer is employed).   Price M. **95.**—

Directions for use. The ends of the conductor X of which the resistance is to be determined are coupled to the terminals D and E, the battery S (2—3 cells) to A and B. The galvanometer G to J² and J¹ turning the instrument until it is in the magnetic meridian, in other words, until the needle remains at zero. From the four comparative resistances of **1, 10, 100, 1000,***) that one should be selected which causes the pointer to remain as nearly as possible in the middle of the scale, when the galvanometer is not deflected by making and breaking circuit with key a. The reading of the pointer J has finally to be multiplied by the resistance in circuit, that is by either **1, 10, 100,** or **1000.**

No. 388.
Portable set for resistance measurements.

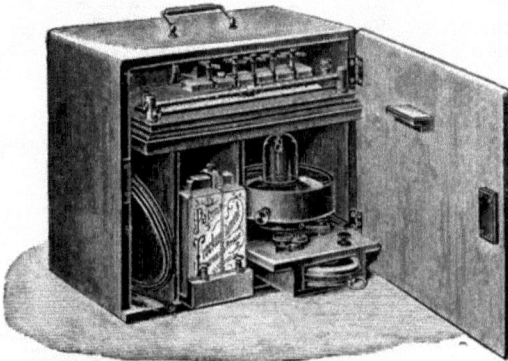

This portable Apparatus for resistance measurements comprises the bridge No. 391, galvanometer No. 366, watch shape telephone No. 1, three dry cells No. 501a coupled in series, all fitting securely in separate compartments in a strong lock-up oak case in which a spare compartment for connecting wires, terminals &c. is provided.
Price M. **275.**—

Total weight approximately 11 Kilo.

*) See foot-note on page 60.

VII. Accessories and Cells.

Keys and Commutators for measuring instruments,

highly finished and mounted on polished mahogany bases.

No. **490. Simple circuit closer,** du Bois-Reymond type, with sliding contact for large currents. Price M. **10.—**

No. **490a. Double successive key,** of similar design. Price M. **30.—**

No. 491.　　M. 10.—.　　　　No. 491a.　　M. 20.—.　　　　No. 491b.　　M. 22.—.

Simple, double, and **double successive keys,** with **platinum contacts,** only intended for small currents; the first two types with an arrangement for permanently closing the circuit.

No. 492 M. 15.—.　　　　No. 492a. M. 28.—.　　　　No. 492b. M. 32.—.

Simple, double, and **double successive keys,** with detacheable **mercury cups** suitable for larger currents; the first two types with an arrangement for permanently closing the circuit.

No. 493.　M. 15.—.　　　　No. 493a.　　M. 45.—.　　　　No. 493b.　　M. 30.—.

Commutators, with **plug, brush,** and **spring contacts;** the first suitable for small currents and the two latter for larger currents.

No. 494.   M. 25.—   No. 494a.   M. 30.—
Revolving Commutator   and   **Rocking Commutator**
the former with 4, the latter with 6 mercury cup contacts.

No. 495.   M. 33.—   No. 495a.   M. 40.—
New Switch   and   **Weber Switch**
the former with 4, the latter with 8 removeable mercury cups.

Battery Selectors.

No. **496. Daurer's Universal-Pachy-**
trop. To enable a current adapted to the
various classes of measurement to be obtained
from a battery without inconvenience or delay,
the use of the above battery selector is advisable, by means of which several cells can be
connected either in series or parallel, or any
single cell may be selected as required. Made
with 4, 5, 6, or 10 pairs of terminals. Price M. **50.**—, **55.**—, **60.**—, or **100.**—

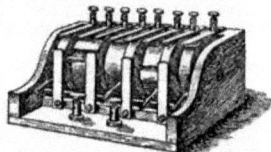

No. 496.

Directions for use: Each cell or combination of cells is connected to the pairs
of terminals fastened to the German-silver strips on the top of the apparatus, care
being taken that the poles of the cell or cells are correctly coupled to the terminals as
marked. The current is taken off from the larger pair of terminals. According to the
position of the levers the following combinations are then at once available:

1. All cells cut out of circuit: every lever in its mid-position.
2. All cells coupled in series: the two outermost levers pushed down and
   all the others pushed up.
3. All cells coupled in parallel: all the levers pushed down.
4. Any single cell: the two levers corresponding thereto pushed down, and
   all the rest in their mid-position.
5. 2 or more cells coupled in series: the two levers between which the cells
   to be connected are mounted are pushed down, and the lever or levers between
   these are pushed up, the remaining levers in their mid-position.
6. 2 or more cells coupled in parallel: the corresponding levers commencing
   from the right hand are pushed down, the remaining levers in their mid-position.

Battery-selectors with **plug contacts** in various designs specially
quoted for on receipt of particulars of requirements.

Keys and Commutators
for capacity and insulation measurements.

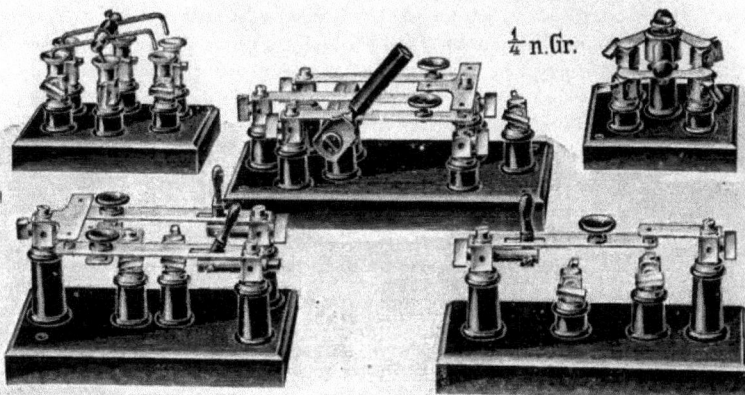

¼ n. Gr.

No. 499a.　　No. 497a.　　　　No. 498.　　　　No. 497.　　　No. 499.

We have, in designing these various **keys** and **commutators**, specially borne in mind the necessity for obtaining as high a degree of insulation as possible between all parts carrying current, and to earth; for this purpose not only the spring contact strips but also the various terminals are all mounted on sufficiently high ebonite pillars which are so arranged and designed as to facilitate access thereto from all sides to enable all dust and moisture to be removed.

No. **497. Charge and discharge key** for capacity measurements in condensers or cables, on Sabine's principle.　　　　　Price M. **30.—**

The spring contact which is to be connected to the cable or terminal of condenser is kept free when the lever is in its backward position; if this is in its mid position the spring contact touches the contact which is to be connected to one pole of the battery (Charge position); if the lever is in its forward position, contact is made with the lower fixed contact which is to be connected to the galvanometer (Discharge position). The fourth terminal is for earth, galvanometer, the second terminal of condenser and the other pole of battery.

No. 497a. Cable key for insulation measurements for reversing battery poles. Price M. **55.—**

Two spring contacts, one of which is connected to earth and the other to the galvanometer, are pressed separately and alternately against an upper contact connected to the —pole of battery and a lower contact connected to the +pole of same: in the latter case the ebonite lever is in its forward position, in the former, in its mid position, if the lever is in its backward position the spring contact remains free between the others.

This key is even more convenient than that previously described for capacity measurements, the method of connection and manipulation remaining as before. For this purpose the free terminal is again connected to earth, galvanometer, condenser and battery, the cable to one spring contact and the remaining terminal of condenser to the other.

No. **498. Compensation-key** for capacity measurements by Thomson's (Lord Kelvin's) method. Price M. **90.—**

One terminal of each of the two condensers which are to be compared is connected to each of the two outside spring contacts, one of the galvanometer leads is taken to the centre spring contact whilst the other together with the remaining condenser terminals are coupled to the earth terminal. Charge and discharge are effected by means of the lever at side, in its mid position all three springs are free from contacts, when turned backwards the outer springs make contact with the battery contacts for charging; when turned forwards the outer springs first make contact with a metal bridge, thus the condensers discharge against each other; on further turning, the centre spring contact also comes against the bridging piece and any residual charge is thereby discharged through the galvanometer to earth.

The series rheostats, page 41, having branch plugs at side, are especially suitable for use as variable resistances through which the battery is placed in circuit.

No. **499. Spring commutator** for insulation measurements, with four fixed terminals which can be coupled in pairs or alternately through two spring contacts mounted on an ebonite handle to turn as required, similar in principle to the revolving commutator No. 494. Price M. **40.—**

No. **499a. Switch** with six mercury cup contacts, mounted on ebonite pillars, resembling the rocking-commutator No. 494a. Price M. **55.—**

Complete sets of apparatus for insulation and capacity measurements, mounted on marble bases, with all connections ready for use can be supplied; the price depending on the apparatus required.

Dry cells for measuring purposes.

$^1/_3$ n. Gr.

These **dry cells** are especially suitable for measuring purposes on account of their high electromotive force, but, like other dry cells, they are not designed to give large currents, and it is therefore adviseable to protect them from short circuit or discharge by a very small resistance; the internal resistance and electromotive force in these as in other dry cells is not absolutely constant, but when the cells are exhausted they can be repeatedly recharged; they are supplied in the two following types.

No. **501a** in strong zinc case, rectangular 8×4×12 cm, 15,5 cm high including terminal; weight 1 Kilo.   Price each M. **2.75**

No. **501c** in insulating case, square section 3×3×7 cm, 9 cm high including terminal; weight 120 grammes.   Price each M. **1.50**

Directions for recharging: The corresponding poles of the cell to be recharged and the source of current are connected together; and to obtain the best results, the cells should be regenerated as soon as the E. M. F. falls below 1 volt, without waiting until they are almost entirely run down. No. 501a should be charged at about 0.4 ampères and No. 501c at about 0.2 ampères for a few hours.

Portable batteries in wood cases.

¹/₅ n. Gr.

Portable batteries, in strong oak cases for measuring purposes. Cells of No. 501a type, not exceeding 30 in number, and of No. 501c not exceeding 100, are securely mounted for travelling purposes in case, fitted if desired with commutator, selector, &c. Where space is available the larger size cells are preferable, as these, with a low internal resistance, vary but slightly in voltage in the course of time. The smaller cells are especially suitable for use with portable apparatus for determining insulation, as in this case it is not so much a question of the internal resistance of the battery as of obtaining the highest possible potential within the smallest space.

No. **502. Battery** of 30 cells No. 501a, coupled in series, in oak case, with compartments, and selector for 10, 20, or the total number of cells. Price M. **150.**

Other combinations quoted for on receipt of particulars.

VIII.

Instruments
for magnetic measurements.

No. **410.** **Portable Bifilar-Variometer for terrestrial mag-netism;** Kohlrausch type. This type differs from the Gauss pattern by the use of a small tubular magnet which consequently causes but slight magnetic disturbances in its immediate neighbourhood. The bifilar suspension consists of two extremely fine brass wires (0.05 mm dia.) 30 cm long and about 8 mm apart. The magnet is powerfully damped, and to ensure its permanency it is several times exposed to a high temperature and for long periods in accordance with the Barus Strouhal method. The variometer is easily mounted and its constants determined by the torsion head which is fitted with vernier and micrometer screws. To control the invariability of the suspension an adjustable mirror is mounted on the torsion tube. Magnet, and mirror, capable of being turned, are carefully shielded from air currents, and a thermometer is mounted in the damper . . . . M. **350.**—

(Wiedemann's Annalen XV, 1882, page 553.)

$\frac{1}{4}$ n. Gr.

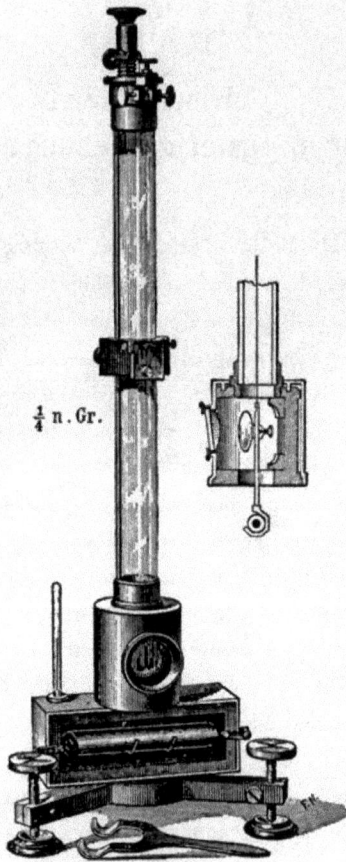

No. 410.
Portable Bifilar – Magnetometer
Kohlrausch type.

ca.
1/4 n. Gr.

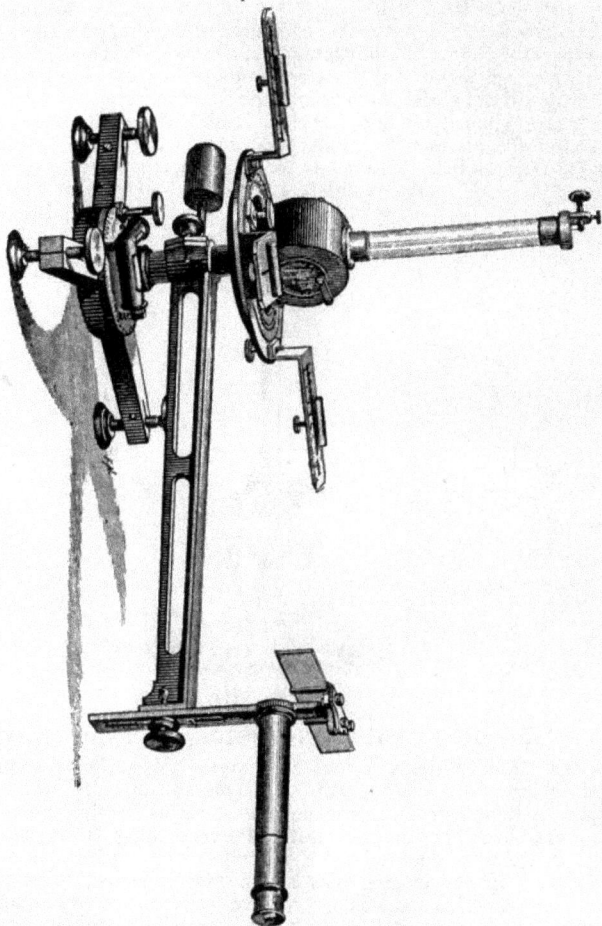

No. 411.
Portable Intensity-Variometer, Kohlrausch type.

No. **411. Intensity-Variometer** for terrestrial magnetism with four controlling magnets: Kohlrausch type. The needle in this instrument is formed of a true plane steel mirror and controlled by four small bar magnets which give a very constant magnetic field in their immediate neigbourhood. It is deflected 90° from the meridian so that declination variations have no effect. As compared with the instrument previously described this offers the advantage that the sensitiveness can be adjusted to any desired amount, and its constants are easily determined by means of the graduated circle above which the framework supporting the magnets revolves. A small reading telescope with opal glass scale is mounted direct on the instrument. This apparatus, which is very portable, can also be used as local variometer to determine the variations of horizontal magnetic intensity between different places with extreme accuracy (1:10000)

Price M. **375.**—

(Wiedemann's Annalen XV, 1882, page 540.)

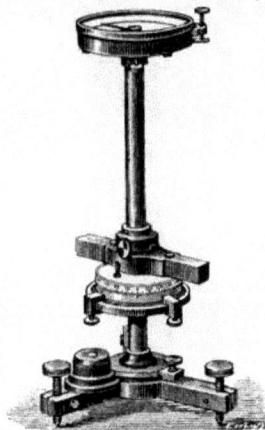

ca. ¹/₄ n. Gr.

No. 411a.

Simple Local-variometer, Kohlrausch type.

No. 411a. Simple Local-variometer to determine local variations in the horizontal intensity. Whilst the method of using and the general design of this instrument remain the same as in the above described variometer No. 411; in this instance every thing is simplified and the entire instrument rendered more portable. The combination of 4 magnets is replaced by one magnet of a suitable form, which, together with the graduated circle is adjustable concentrically below the magnet needle which is here supported on a pivot. Suitable clamping screws ensure the constancy of the turning-angle, and the instrument is sufficiently accurate for most purposes (1 : 1000) . . . . Price M. **200.**—

(Wiedemann's Annalen XXIX, 1886 page 47.)

No. **412.** **Absolute Bifilar-Magnetometer,** Kohlrausch type. In contrast to other magnetometers, the method of suspension here, is by two very fine wires widely separated from each other (12 cm). On account of the accuracy with which these conditions can be determined, the deflections, when the magnet is reversed, give the product of the bar magnetism and the terrestrial magnetism in absolute measure. Simultaneous observations of a unifilar magnetometer (see No. 415 and 415a) deflected by the magnet give the earth's magnetic force in absolute measure. This method is preferable because no time observations are required and neither temperature nor locally induced magnetism can affect the result.

The apparatus comprises the suspension with divided circle, tubular magnet with rotatable mirror, damper for use with fluids, and vibration chamber. **Price M. 230.—**

(Wiedemann's Annalen XVII, 1882, page 737.)

No. **413.** **Absolute Bifilar-Galvanometer,** Kohlrausch type. A wire ring 20 cm diameter with a large number of turns is bifilar suspended in a similar manner to the previous instrument. Its deflection by means of a current gives the product of the current strength, area of coil and terrestrial magnetism. In combination with a tangent galvanometer (see Nos. 374 to 376) or with a magnetometer (see Nos. 415 and 415a) it can be employed either for absolute measurement of the current or of the terrestrial magnetism.

Suspension, wire ring with rotatable mirror and vibration chamber. **Price M. 250.—**

(Wiedemann's Annalen XVII, 1882, page 737.)

No. **415.** **Unifilar Magnetometer,** Kohlrausch type, with a small magnet mounted at back of mirror, which, working in a very narrow chamber with inserted vanes, acts as an efficient air damper. **Price M. 120.—**

No. **415a.** **Unifilar Magnetometer,** Kohlrausch type, later pattern, with ring magnet, in copper damper; the mirror adjustable against the magnet and the entire system shielded from external air currents by a wooden case. **Price M. 180.—**

No. **420.** **Compensation Magnetometer,** Weber and Kohlrausch type, with stand and vibration chamber. **Price M. 300.—**

No. **421.** Gauss-Weber **Magnetometer,** for mirror readings, to determine the horizontal component of the earth's magnetic force, and also the declination; with magnets 8 cm long and rails for same 50 cm long, copper damper, vibration chamber and all other requisite accessories. **Price M. 400.—**

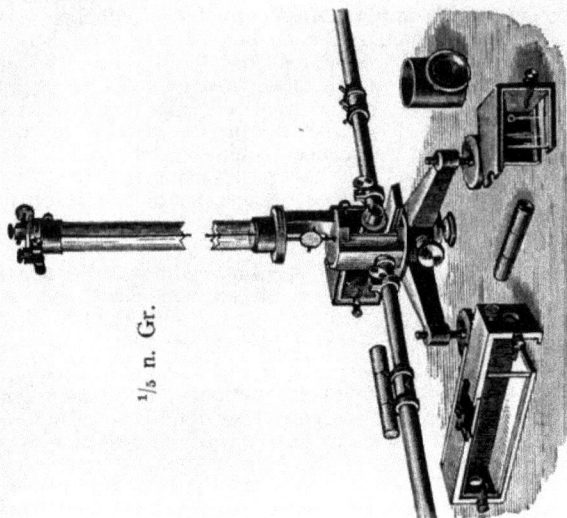

1/5 n. Gr.

No. 421.
Gauss–Weber Magnetometer.

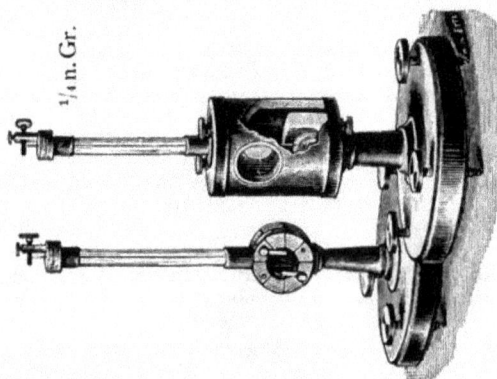

1/4 n. Gr.

No. 415 and 415a.
Magnetometer, Kohlrausch type.

No. 425.
Earth Inductor, W. Weber's type.

$^1/_8$ n. Gr.

No. **425**. **Earth Inductor**, W. Weber's type, for measurement of terrestrial magnetic inclination, with induction ring 20 cm diameter, very strongly built and fitted with convenient arrangements for adjustment. The mirror magnetometer which is included in the apparatus for the purpose of adjusting it to the magnetic meridian renders it also available as a tangent galvanometer . . . . Price M. **650.**—

No. **425a**. **As above,** with induction ring 40 cm dia. Price M. **850.**—

No. **427**. **Astatic Reflecting galvanometer***) for use with the earth inductors Nos. 425 and 425a with corresponding vibration periodicity and extreme sensitiveness and which, by means of an easily inserted copper damper and double winding is available for galvanic work.

Price M. **400.**—

----

*) See page 19.

No. 426.
Differential Earth Inductor. L. Weber's type.

$\frac{1}{8}$ n. Gr.

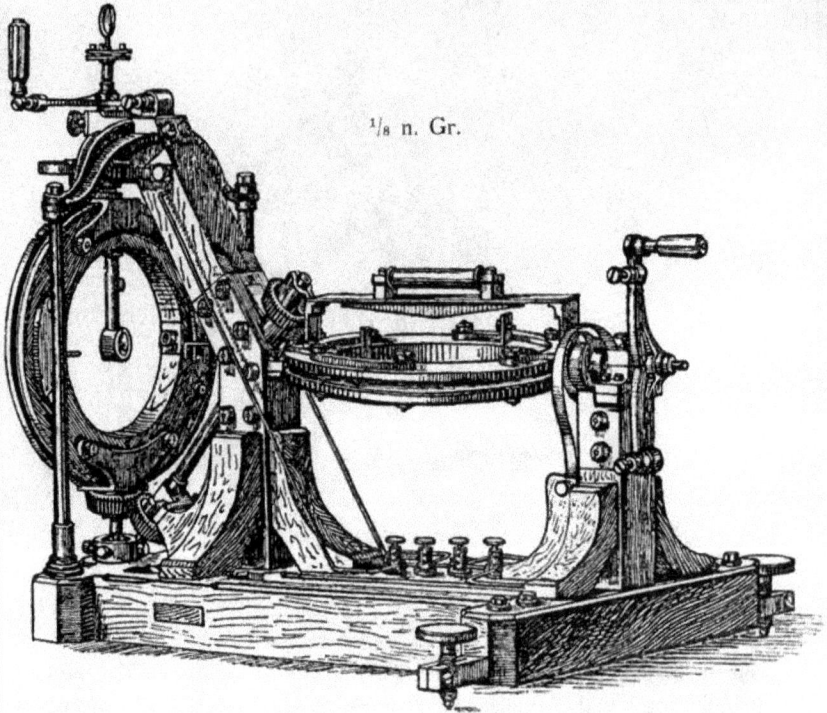

No. **426. Differential Earth Inductor,** L. Weber's type, for determining rapidly the angle of inclination. Two coils, as near as possible duplicates of each other and interchangeable in their bearings, with their axes at right angles to each other are revolved together. The measurement is made with a differential galvanometer and occupies scarcely one minute. The strength of current is equalised in both coils by inserting resistances. The tangent of the angle of deflection is then equal to the ratio of the resistances of both circuits. Price M. **2200.—**

(Sitzungs-Berichte der Akademie der Wissenschaften, Berlin, XLIX, December 1885.)

No. 431.
Large ☁ Electromagnet.

$^1/_{10}$ n. Gr.

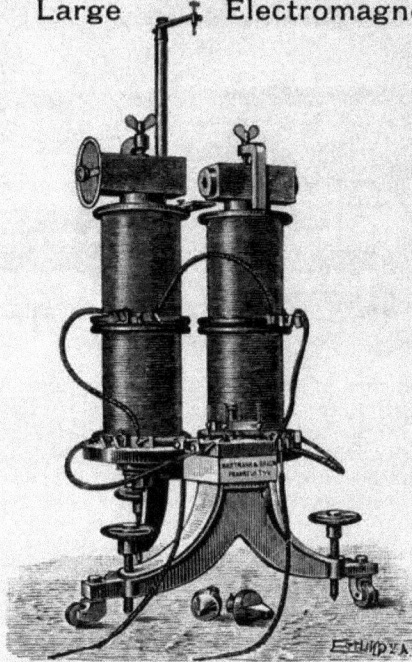

No. **431.** **Large Electromagnet,** made from Swedish charcoal iron, with vertical limbs 40 cm high and 70 mm in diameter, pole pieces of square-section, with hole bored through; they can be adjusted and clamped in any desired position on the surfaces of the limbs. The winding, of 3 mm copper wire, is in four removable sections on brass forms and each form has suitable terminals mounted thereon to enable the windings to be easily connected in series or parallel as required. The stand is fitted with rollers for convenience in moving the instrument, and screws are provided to fix it in any desired position. A commutator and vertical pillar with a small adjustable table are also mounted on the stand, and a small piece of bismuth and a glass trough for diamagnetic experiments are sent with the apparatus. Three different pairs of insets for the pole pieces in various well known forms are also included. Price M. **1400.**—

Extra apparatus:

Waltenhofen **Pendulum** for experiments in induction. Price M. **255.**—

Polarisation Apparatus for experiments in diamagnetism; fitting into the holes bored in pole pieces . . . . Price M. **110.**—

Parallelepipedon with partially silvered surfaces, made in Faraday glass (silicate and borate of lead) . . . . Price M. **30.**—

No. 504. Bismuth Spiral
for measurements of magnetic fields.
Lenard's type.

¹/₃ n. Gr.

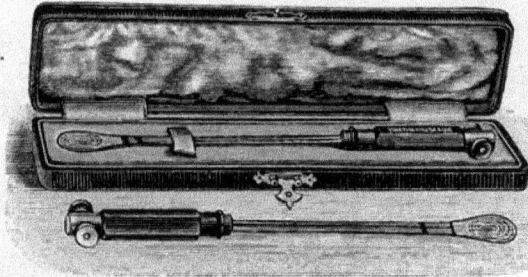

Measurements of the intensity of magnetic fields by means of Lenard's bismuth spiral are effected through the change of resistance which occurs in bismuth when in a magnetic field. In this instrument a thin wire of chemically pure bismuth, well insulated, is double wound as a flat spiral, and the ends soldered to two flat copper strips which are fitted with requisite terminals and mounted in an ebonite handle; the spiral is cemented between two thin mica discs to protect it from damage. As the entire thickness of the spiral is only about 1 mm, the instrument can be used in very confined spaces, as for instance in the clearance space between pole pieces and armature of a dynamo. The alteration in resistance affords the means for determining the number of lines of force in the field tested, 1000 representing approximately 5% alteration in the resistance; the calibration curve which is supplied with each instrument gives the relative proportion with greater accuracy. Price M. 50.—

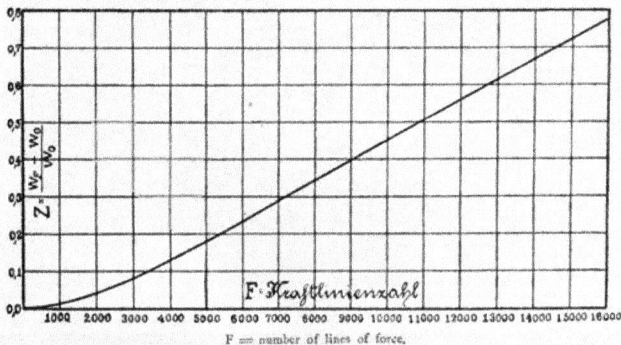

F = number of lines of force.

Reduced copy of a calibration curve which enables the relation between the increase in resistance of the bismuth spiral and the number of lines of force in the field under examination to be determined direct. The ordinate values represent the increase in resistance Z, determined from the resistance of the bismuth spiral Wo in field zero and WF in field F.

No. 560. Simple apparatus
for investigating the magnetic properties of iron
by means of the bismuth spiral.

⅓n.Gr

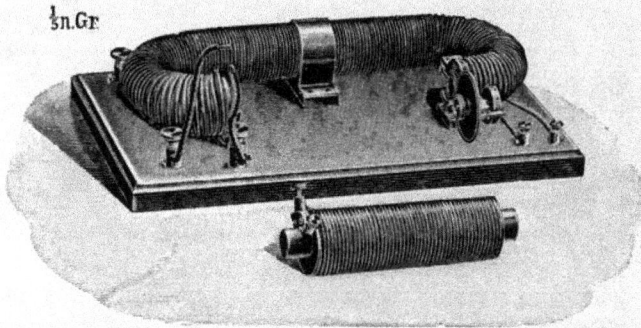

No. 560. This **apparatus for investigating the magnetic properties of various classes of iron** consists of an oval, wire wound electromagnet; one side is easily removable and the piece of iron to be tested can be inserted in the space thus provided in combination with a bismuth spiral No. 504 fixed in the narrow air space then remaining; by means of a micrometer attachment mounted on the apparatus this air space can be exactly determined. The winding is designed to give comparatively strong fields with small currents.   Price M. **210.—**

Instructions for use: The iron to be tested is adjusted as accurately as possible in size to the standard removable bar and the intensity of the field is determined by means of the bismuth spiral for both the standard bar and the bar under examination using the same number of ampère turns in each case. These two results give the comparative values of both the standard and the testbars, and, if the magnetic properties of the standard bar have been plotted as a curve, those of the bar under examination can easily be plotted and compared therewith.

Galvanometer No. 335 page 14 is specially adapted for this instrument as it is not affected by other electromagnets in its proximity.

The following arrangement, using a bridge, is very convenient for determining the intensity of magnetic fields.

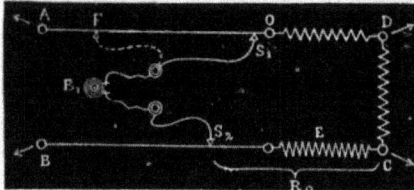

Two resistances, of 1 ohm each, OD and DC, are connected to a stretched measuring wire fitted with sliding contact S_1 also a resistance E equal to that of the bismuth spiral at its lowest temperature and lastly a second measuring wire with sliding contact S_2. The spiral is connected to S_1 and S_2, the galvanometer to A and C, and the battery to B and D. The bridge is then balanced with the spiral in field zero by placing the sliding contact S_1 on the zero of the measuring wire and adjusting S_2 until balance is obtained, then, as OD = DC, DS_2 = WO, i. e. equal the resistance of the spiral at the prevailing temperature. Without moving S_2, S_1 is then adjusted with the spiral in the field to be determined; then OF $= \dfrac{Wf - Wo}{Wo}$

i. e. is equal to the increase in resistance Z of the spiral. The measuring wire can be scaled to give this quotient direct, and, provided the same spiral is always employed, can also be scaled to give the intensity of the field direct.

No. 560a.
Complete apparatus for investigating the magnetic properties of iron.

⅛ n. Gr.

No. 560a. Complete apparatus for investigating the magnetic properties of iron. The apparatus illustrated herewith is especially suitable for use where it is desirable to make the test as quickly and easily as possible, without having first to connect for this special purpose the various instruments perhaps existing for accurate resistance measurements; it is extremely useful for factories, works, such as foundries, iron works &c., where a staff competent to carry out electrical testing is not maintained. It comprises the electromagnet No. 560 with bismuth spiral, a switch for heavy currents and an ammeter, unaffected by neighbouring magnetic fields, to measure the current employed in exciting the electromagnet, a galvanometer and double key for the measuring current, and lastly a specially arranged bridge on which the number of lines of force can be read off direct. All the instruments are mounted on a mahogany base board ready for use and so arranged that, using the instructions printed below, comparative tests of various samples can be made without any previous training.   Price M. 585.—

Instruction for use: A test piece of the iron to be examined, as nearly as possible the same size as the standard (a cylinder 200 mm long and 25 mm diameter), is prepared and, encircled by the moveable wire coil, placed between the poles of the electromagnet with its left hand face bearing firmly against the magnet pole. An adjustable source of current, preferably accumulators capable of discharging up to a rate of 15 ampères, is coupled to the large terminals. The testing battery of 2 or 3 cells is connected to terminals b. The sliding contact S_1 of the front measuring wire is then placed on the zero mark of its scale and contact for the measuring current made by depressing the key t, adjusting the sliding contact S_2 of the measuring wire at back and repeatedly making and breaking contact with t until the galvanometer is not deflected. The exciting current is then switched on by switch T, measured by the ammeter, and the measuring battery again brought in circuit by t; then leaving S_2 in its present position on the measuring wire at back, S_1 on the front measuring wire is adjusted until the galvanometer again is not deflected. T is then at once switched off to avoid heating the electromagnet and the number of lines of force read off direct on the scale at S_1.

It is important that the amount of air space when using test pieces of various lengths shall be accurately known and this is effected by turning the milled edge of the disc forwards and reading off the graduations. More complete instructions will be furnished with the apparatus if desired.

IX.
Optical Apparatus.

Mirrors.

Steel magnet mirrors of the highest finish and true planes.

Diameter in mm	Polished on one side	Polished on both sides
15—20	M. 20.—	M. 45.—
21—25	M. 25.—	M. 55.—
26—30	M. 33.—	M. 70.—

If desired magnets of other shapes can be supplied with circular polished reflecting surfaces.

Thin true plane mirrors, absolutely accurate, silvered.

Diameter in mm	Thickness 0.2—0.4 mm.	Thickness 0.5—1 mm.
5—10	M. 6.—	M. 4.—
11—15	M. 9.—	M. 7.—
16—20	M. 12.—	M. 10.—
21—25	M. 18.—	M. 15.—
26—30	M. 24.—	M. 20.—

True plane mirrors from 2—5 mm in thickness,

round . . . . per square centimeter M. **1.20**

square . . . . „ „ „ „ **1.40**

assuming the area to be not less than 6 square centimeters.

Special quotations will be given for mirrors smaller than those above listed, larger than 50 square centimeters, or of greater thickness than **7** mm.

The price of unsilvered mirrors is 10% less.

Concave mirrors for measuring instruments, focus 50 cm, 0.4 mm in thickness, 10, 15 and 20 mm in diameter and silvered at back.

Price M. **6.—, 9.—** and **12.—**

Concave mirrors for reflecting telescopes in glass with polished silvered surface, or in hard silver coloured bronze alloy, from 5 to 20 cm diameter, will be specially quoted for.

Eye Pieces.

Terrestrial Eye Piece Lux

Mittenzwey type.

The **terrestrial Lux Eye piece** is formed with a bi-convex and a plano-convex lens, both achromatic, and covers a large field with perfect definition throughout. It is well adapted for direct use as a magnifying glass, the great distance from the object being very advantageous.

Equivalent focus . . . cm	3	2.5	2	1.5	1.25	1	0.8
Length . . . . . about cm	18	15	12	9	7.5	6	5
Price, including mountings, M.	50.—	42.—	36.—	30.—	30.—	30.—	30.—

Astronomical Eye Pieces.

Improved	Holosteric
Huyghen Eye Piece	Micrometer Eye Piece

Mittenzwey type.

The **Improved Huyghen Eye piece** is formed with two simple lenses mounted at the requisite distance apart to ensure achromatism; it covers a field of about 55° with perfect definition throughout, excelling in this respect the excellent three lens Steinheil eye pieces; and is of great service as long focus eye piece for comet seeking.

Equivalent focus . . cm	6	5	4	3	2	1.5	1	0.75	0.5
Price, mounted in cells*) M.	30.—	24.—	18.—	15.—	12.—	10.—	10.—	10.—	12.—

The **Holosteric Micrometer Eye piece** is formed with two cemented lenses, and therefore has only two refractions, glass-air, it is quite free from flare, and gives perfectly sharp flat pictures without aberration in a field of 35°. It is especially noticeable for the great distance from the object at which it works, 16 mm in an eye piece of 2 cm focus, and is therefore extremely useful as an aplanatic magnifying glass.

Equivalent focus . . cm	3	2.5	2	1.5	1.25	1
Price, mounted in cells*) M.	30.—	25.—	20.—	18.—	18.—	18.—

*) If these eye pieces are to be fitted to an existing telescope we desire the eye piece mount or a flange to be sent to us and the cost of fitting is extra, The outer tube for the holosteric and the euryscopic-aplanatic micrometer eye pieces on page **82** are not included in the prices quoted, although shown In the engravings to illustrate the great distance between lens and object.

Euryscopic-aplanatic Micrometer Eye piece
Mittenzwey type.

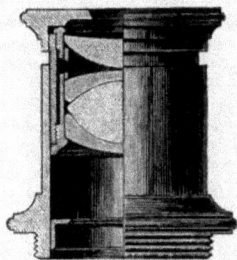

The **Euryscopic Micrometer Eye piece** is formed from a cemented three lens combination strongly over-corrected for spherical and chromatic errors and a concavo-convex lens having similar errors in the opposite direction. This system complies to a very high degree with all the essentials of a perfect micrometer eye piece: the euryscopic and orthoscopic stability of the achromatism, aplanatism both in and without the axis, long distance between lens and object, and freedom from troublesome reflections render it as perfect as possible. In a field of view of 53° the distance between lens and object in an eye piece of 2 cm focus is 11.5 mm.

Equivalent focus . . cm	2	1.75	1.5	1.25	1	0.75	0.5
Price, mounted in cells M. (see foot note on page 81)	25.—	23.—	20.—	20.—	20.—	20.—	20.—

Eye piece cover with small aperture to the micrometer eye pieces, for measurements with bright threads on a dark ground (according to Abbe) . . . Price M. 1.50

Ramsden astronomical Eye pieces (2 plano-convex lenses in brass mounts, image in front of the system) of equiv. foci from 0.5 to 2 cm . . . . Price M. 9.—

Steinheil astronomical Eye pieces (3 plano-convex lenses in brass mounts with screw type adjustment on to the cross threads, image between the collective and second lenses) of equiv. foci from 0.5 to 2.5 cm . . . . Price M. 12.—
of equiv. foci of 3.0, 4.5 and 6.0 cm Price M. 24.—, 40.— and 60.—

Accurately centered **lenses** of any desired focus, plano-convex or bi-convex, plano-concave or biconcave. Prices according to quantities ordered.

Achromatic Object glasses for telescopes
made up of two lenses.

Clear Aperture in mm	Focus in cm	Price including mount M	Clear Aperture in mm	Focus in cm	Price including mount M
10	8, 10, 12	6.—	35	30, 32, 40	24.—
15	8, 10, 12	8.—	40	32, 40, 48	32.—
20	8, 10, 15, 20	11.—	50	50, 55, 60	45.—
25	18, 20, 25, 30	14.—	55	55, 60	55.—
27	18, 20, 25, 30	16.—	60	60, 72	70.—
30	25, 30	18.—	70	70, 84	115.—

Larger object glasses with aperture ratio of 1:12 or 1:15. Prices to quote.
Powerful achromatic object glasses of **very short focus** made up of three lenses.

Clear aperture in mm	20	25	27	30	40	50	60
Focus . . . . „ cm	6	7.5	8	9	12	15	18
Price . . . . „ M.	16.—	21.—	24.—	30.—	45.—	70.—	100.—

Prices will be quoted for mounting object glasses in wood, brass or steel tubes.

Prisms.

Prisms in **crown** or **flint glass** with point angle of 60° or any smaller angle preferred, with perfect true plane polished surfaces.

Clear aperture or length of side in mm	With **2** polished **round** surfaces	With **3** polished **round** surfaces	With **2** polished **quadratic*)** surfaces	With **3** polished **quadratic*)** surfaces
10	—	—	M. 6.—	M. 8.—
15	—	—	„ 7.—	„ 9.—
20	M. 7.—	M. 9.—	„ 9.—	„ 12.—
25	„ 10.—	„ 13.—	„ 13.—	„ 17.—
30	„ 14.—	„ 18.—	„ 18.—	„ 24.—
35	„ 19.—	„ 25.—	„ 25.—	„ 33.—
40	„ 25 .—	„ 33.—	„ 33.—	„ 44.—
45	„ 32.—	„ 43.—	„ 43.—	„ 56.—
50	„ 40.—	„ 54.—	„ 54.—	„ 72.—
60	„ 60.—	„ 80.—	„ 80.—	„ 106.—
70	„ 80.—	„ 106.—	„ 106.—	„ 140.—
80	„ 100.—	„ 130.—	„ 130.—	„ 170.—

*) Prisms with rectangular surfaces will be specially charged for.

Prisms in quarz, thallium glass, Feil's and Chance's double extra dense flint glass, or the new and valuable series of Jena glasses, such as phosphate-crown, barium-phosphate and borate-crown, borate-flint and silicate-flint (with n D = 1.9626) are supplied as cheaply as possible, the cost varying with the material employed.

Rectangular reflection prisms in **crown glass,** perfectly true in angles and surfaces and free from pyramidal error.

Clear aperture or length of side in mm	With **round** Catheten surfaces	With **rectangular** Catheten surfaces	Clear aperture or length of side in mm	With **round** surfaces	With **rectangular** surfaces
3—8	—	M. 6.—**)	45	M. 63.—	M. 95.—
10	—	„ 8.—**)	50	„ 80.—	„ 120.—
15	—	„ 12.—**)	60	„ 100.—	„ 150.—
20	M. 21.—	„ 27.—	70	„ 150.—	—
25	„ 26.—	„ 35.—	80	„ 240.—	—
30	„ 32.—	„ 45.—	90	„ 350.—	—
35	„ 40.—	„ 60.—	100	„ 500.—	—
40	„ 50.—	„ 75.—	120	„ 800.—	—

**) The reflection prisms from 3 to 15 mm in length of side are only intended for lighting purposes and are not therefore made with absolutely accurate angles and surfaces.

Silvering the hypothenuse surface of the above reflection prisms per □cm M. 0.20
with a minimum charge of „ 1.—

Straight sight prisms, C. Braun's type.

Aperture in mm .	8	10	15	20	25	30
of 3 Prisms . M.	15.—	20.—	30.—	40.—	55.—	75.—
of 5 Prisms . M.	20.—	30.—	45.—	65.—	90.—	120.—

Prisms for various purposes:

Prisms for fluids, Steinheil pattern, with air tight true plane discs, bore 20 and 30 mm
M. 80.— and M. 140.—

Angle prisms in holder and case, length of side 20 and 25 mm „ 9.— „ „ 12.—

Combined prisms of every description in exact accordance with any specification will be quoted for.

1/3 n. Gr.

No. 446. Spectrometer.

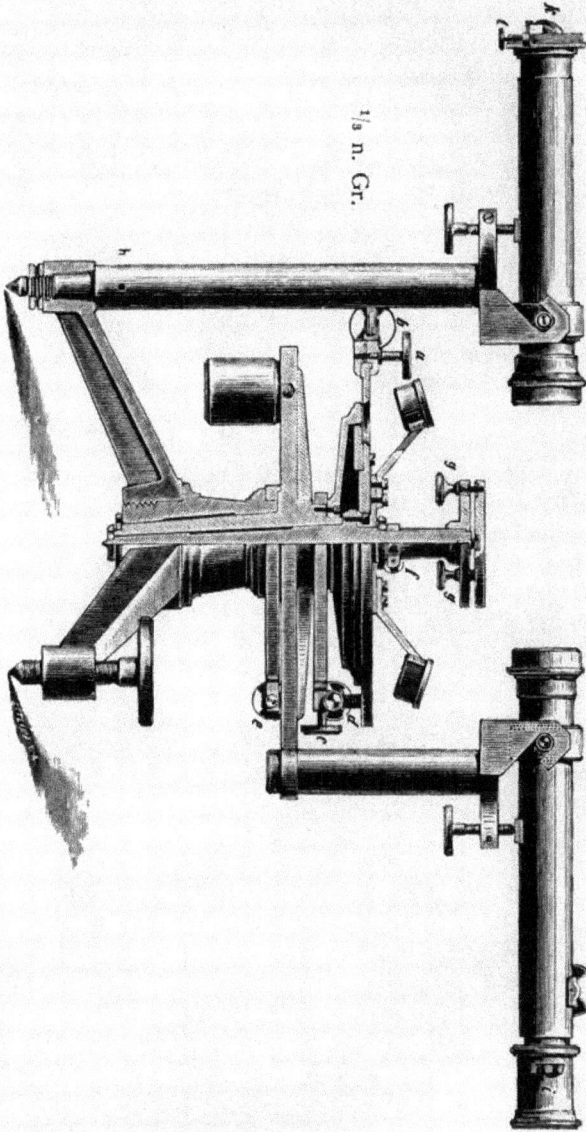

Spectrometer.

No. **446. Spectrometer** with covered circle 12 cm diameter. Reading to 30 sec. by two verniers and magnifying glasses; circle and observing tube are independant of each other, can be rotated, and are fitted with micrometer adjustment; the eye draw of observing tube is adjustable by rack and pinion and the eye piece is fitted with a total reflection prism in lieu of the Gauss mirror eye piece; this can be easily moved to one side after adjustment thereby enabling the position of the telescope to be frequently checked without difficulty whilst working. The slot is fitted with a comparison prism, both observing tube and slot tube have object glasses 25 mm diameter and can be set to a very acute angle with each other, are easily removed and quickly adjusted. The small prism table can be regulated for height and also to the minimum of deflection through frictional contact by hand adjustment.   Price M. **600.—**

Crystal Holder with centreing attachment taking the place of the small prism table and enabling the spectrometer to be used as a goniometer. Price M. **50.—**

No. **454. Kohlrausch Total Reflectometer** for determining the light refracting powers of solid bodies. The substance to be examined can be either transparent or opaque and with either single or double refraction and need only have **one** small plane surface. The instrument has a circle of 10 cm diameter divided into degrees reading on the verniers of the inset alhidade to 3 min. Crystal holder with double ball and socket joints, clamp with disc points, direction sector, thermometer, telescope with cross threads, glass micrometer, a second observing tube with half eye lens and cross engraved in glass in lieu of the object glass, and black glass mirror for observations with polarised light, shield carrier and shield.   Price M. **180.—**

No. **454a. Total Reflectometer,** as above, but with circle 12 cm in diameter, graduated in half degrees, verniers reading to 1 min. on alhidade with micrometric adjustment, magnifying glasses and holders.   Price M. **220.—**

Extras for Nos. 444 and 454a:

1) **Crystal holder** with graduated circle and verniers, and external micrometer screw adjustment . . . . Price M. **45.—**

2) **Nicol's Prism** mounted in small tube fitting the telescope support.
          Price M. **20.—**

3) **Trough** in flint glass with parallel glass sides, for determining the refraction ratios of fluids . . . . Price M. **15.—**

4) **Wood case** . . . .   „   „ **20.—**

(Wiedemann's Annalen IV, 1879, page 1.)

No. 454. Total-Reflectometer (Kohlrausch).

¹/₃ n. Gr.

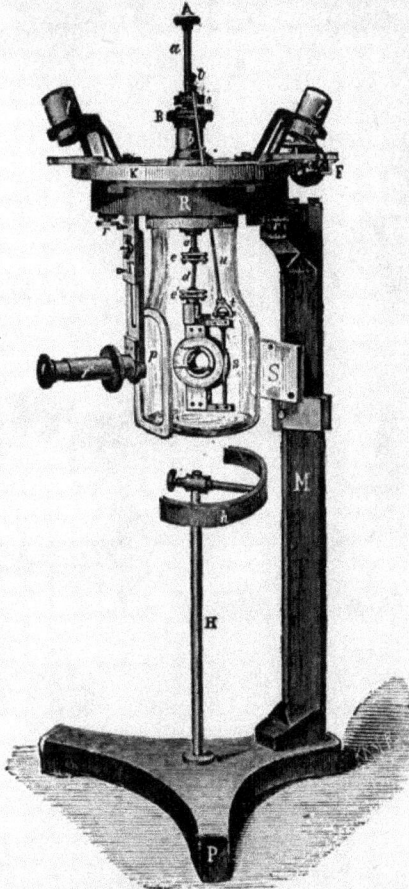

No. **456. Small Spherometer** with three feet and contact screw, one plane and one spherical glass, suitable for practical instruction.   Price M. **55.**—

No. **466. Small Cathetometer,** iron tripod with adjusting screws, steel cylinder 60 cm long with inset scale on German silver. Telescope with rack and pinion adjustment to eye piece, and two interchangeable object glasses for large and extremely small distances; with micrometric adjustment for height.   Price M. **350.**—

No. **477. Small Theodolite,** for physical laboratory work, no iron or steel employed. Diameter of the horizontal circle 12 cm, of the vertical circle 10 cm, reading to 30 sec. Telescope aperture 25 mm, 20 cm focus, eye piece prism, and sun glass.
          Price M. **450.**—

Photographs of Nos. **456, 466** und **477** are at our clients disposal.

No. 387.
Opal glass Photometer.
L. Weber's type.

ca. $^1/_6$ n. Gr.

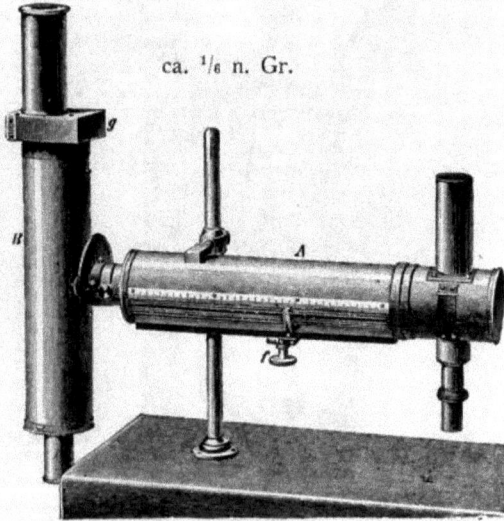

No. **387.** **Weber's Opal glass Photometer** in the portable arrangement here illustrated can be used like the Bunsen Photometer to determine the intensity of a source of light, especially of arc and incandescent lamps, and also, without any extra apparatus, to determine the alteration of light emission in these lamps at different angles of elevation, it is also adapted to determine the degree of heat in incandescent lamps (ratio of the green to the red rays). The chief merit of the instrument however is its ability to measure the diffused light in a room illuminated by these lamps without difficulty and in a very simple manner, for which purpose its extreme portability is a great advantage; it gives, direct in candles per square metre, the degree of illumination on various surfaces in a room.

The standards for comparison are various opal glass discs illuminated by a benzine lamp fitted with means for measuring the height of flame — the constants for these discs are determined beforehand and sent with the apparatus, but can be checked at any time by the instrument itself. The instrument is furnished with Lummer-Brodhun Prisms.   Price M. **420.**—

No. 570.

Standard Photometer with accessories and Lummer-Brodhun Prisms.

No. **570.** The **Standard Photometer** comprises two metal tubes cased with ebonite, over two metres long, these, by means of two supports fitted with adjustable feet, are combined to form a very firm pair of rails, which, through the addition of an elongation piece and support can, if desired, be extended to a total length of somewhat over three metres. The front rail is graduated in half centimetres with the part between 75 and 175 divided into millimetres, and the backrail can, if desired, be graduated to read direct the ratio of intensity of the lights under examination. On these rails are three moveable carriages with tubular clamps, two of which have rack and pinion adjustment to the tubes. The carriage in the centre is fitted with a reversible photometer case with Lummer-Brodhun prisms arranged for the disappearance of one field within the other (equality), another carriage is fitted with a standard amyl-acetate lamp and indicator for height of flame (Hefner light = 0.88 english sperm candle) whilst in the third carriage the various accessories supplied for holding the source of light (candle holder, incandescent lamp socket, table for paraffine lamp &c.) can be inserted.

As above described with rails 2 metres in length . . . . . . Price M. 580.—

Extras:

1) **Extra Rails** & support to form bench of 3 metres . . . . . . Price M. 70.—
2) **Light ratios** marked on back rail . . . . „ „ 25.—
3) **Photometer screen with Lummer-Brodhun Prisms**, for contrast work (two fields appearing equal within the frame of a third) suitable also for equality work.
Price M. 160.—

If this screen is selected in lieu of the other, the extra cost is M. 30.— the latter being charged at M. 130.—.

4) **Graduated arc** to photometer screen, and **Shadow caster** for photometric work at any desired emission angle . . . . Price M. 55.—
5) **Mirror**, adjustable in all planes, with divided circle and graduated arc „ „ 175.—
6) **Dispersion lenses**, 3 of various foci, for reducing the intensity of powerful sources of light, with stand and screen, a true plane glass to equalise the absorption of light in the material of the lens is also included . . . . Price M. 65.—
7) **Incandescent lamp holder** with horizontal and vertical movements and graduated circles to determine amount of same, for photometric measurements with the lamp at any desired angle . . . . Price M. 190.—
8) **Standard Hefner Lamp** (duplicate) with flame measurer, standard gauge, and spare wick tube, with certificate from the Imperial Phys.-Techn. Laboratory. Price M. 50.—
9) **Petroleum lamp**, for comparison of more powerful sources of light. „ „ 30.—
10) **Amyl-acetate, chemically pure**, at cost . . . . Price per Kilo M. 4 to 5.—

No. **571. Photometer** of a similar but simpler construction, specially suitable for incandescent lamp measurements, with 2 rails of angle-iron, 3 m long, one of these with cm divisions. The centre stand is fitted with a small Photometer screen Lummer-Brodhun system for equal fields, one of the others carries a Standard Hefner lamp, the other one is intended for the sources of light to be compared.   Price M. 300.—

No. 570. Apparatus for measuring light. (Standard Photometer.)

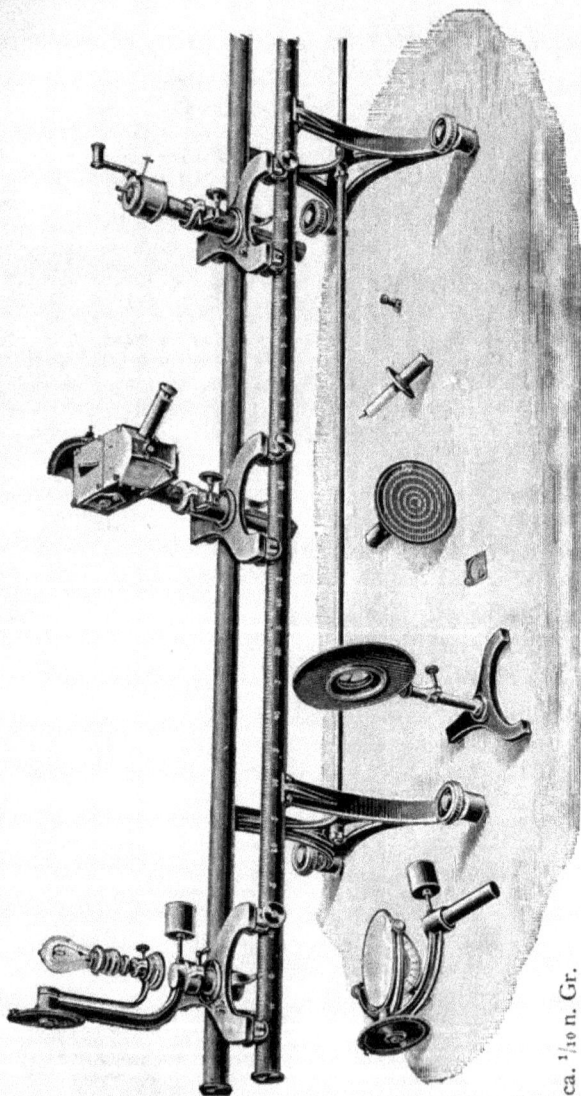

ca. $\frac{1}{10}$ n. Gr.

No. 577. Simple Incandescent Lamp Photometer.

iön.Gr.

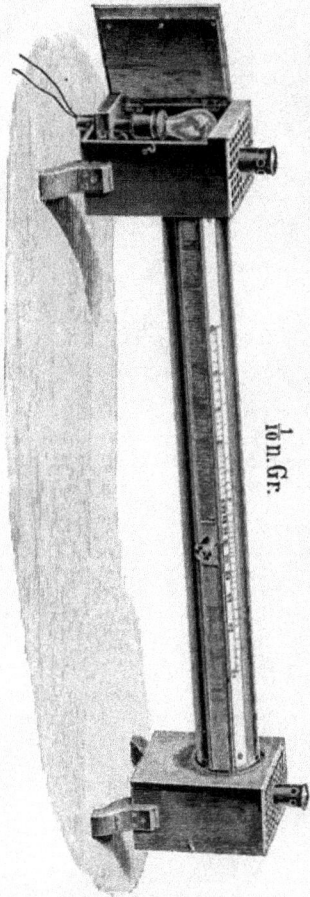

No. 577. Simple Incandescent lamp Photometer, with comparison apparatus on the Joly principle, which enables a sufficient accurate observation to be obtained. The Photometer, 1 metre long, can be used without a dark room for comparing the intensity of light in incandescent lamps[*]) the ratio being read off direct on a graduated scale. The range is from one to ten fold on both sides and the lamps under test can be changed very quickly. Calibrated incandescent lamps serve as standards. Within the ranges cited the absolute amount of light can be determined by comparison with a standard paraffine candle, for which purpose a candle holder is supplied with the apparatus as well as the Edison screw lamp holders, though any other type of holder will be fitted if desired.

Price M. 110.—

Standard incandescent lamps for the usual voltages of 5, 8, 10, 16, 25, and 32 candle powers . . each M. 5.—

It is advisable always to have two standard lamps of the same sort, one of which is kept as absolute standard and only employed to occasionally check the standard usually used.

If desired these lamps will be supplied certified by the Imperial Physico-Technical Laboratory at the extra cost of the certificate.

---

[*]) For measurements of this class it is often advisable to determine at the same time the Watts absorbed by the incandescent lamps, for this purpose the Wattmeter No. 610 l, page 118 is especially suitable.

X. Technical Measuring Instruments
for continuous or intermittent control of
electrical installations.

These **technical measuring instruments** are made on the same lines as our scientific instruments and with the same high quality of material and workmanship throughout. Even externally therefore they compare favourably with the bulk of the electrical measuring instruments now on the market, and this comparison is still more to their advantage when those parts usually hidden behind the dial or scale are examined. To attain a frictionless adjustment the pointers are all mounted in hard polished jewels and not in metal bushes.

Quality of material and manufacture.

We have attached great importance to the production of the most suitable scale in all our instruments and the choice of the principle of construction has been partly influenced by our endeavours to obtain a scale of equal divisions throughout, especially suitable for current meters, or for potential meters one with very wide divisions at a certain part of the scale. — (See pages 126 and 127.)

Scales.

In spite of the extra cost due to the high quality of material and the accuracy of workmanship employed, it will be found that our prices, owing to systematized manufacture, compare favourably with those of our competitors.

Price.

The ampère- and voltmeters illustrated on pages 102 and 103 working on the **electro-magnetic** principle are those as yet most employed. To ensure permanent accuracy in their readings, these instruments are, after the magnetic properties of each iron core has been tested, kept in circuit for a considerable length of time. To reduce the influence of external magnetic fields, the working coils are wound with the maximum number of ampère turns admissible, and in the case of voltmeters a sufficiently high inductionless resistance in Constantan is inserted in series, so that the external temperature affects the readings but slightly, the limit of error not exceeding that of ± 1 per cent usually admitted in technical measuring instruments. Accurate readings with these voltmeters for continuous work are attained when they have been in circuit for about 10 minutes.

Rules observed in calibrating and adjusting electromagnetic instruments.

Mounting electromagnetic instruments on Switchboards.

When **mounting** electromagnetic instruments of any description on a switchboard, it is essential that they be placed at a certain distance away from magnets, such as relays and still more so from the dynamo or from leads carrying current, the distance depending on the maximum current carried and whether lead and return cables are side by side or only one cable is likely to affect the readings, in the latter case the instrument should not be nearer to the cable than about 20 cm with a maximum current of 100 ampères, 40 cm for 300 ampères, 50 cm for 500 ampères and about 70 cm for 1000 ampères, whilst in the former case these distances may be reduced to one half. Single leads do not affect the instruments if they are fixed behind the centre of the coils. Errors due to neglect of these precautions frequently occur, despite of the instructions forwarded with each instrument, but they can be remedied by exchanging or replacing these electro-magnetic instruments by the standard ampère- or voltmeters, page 95, or the hot-wire instruments, page 99 all of which are of the same design externally.

Standard instruments for direct current, dead-beat.

The **standard ampère- and voltmeters** with moving coil in a constant magnetic field, for use with direct current only, are, on account of their greater accuracy — at least $\frac{1}{2}$ per cent — most suitable as check instruments in installations where several volt- and ampèremeters are mounted on the switchboard, they are also dead-beat and therefore extremely well adapted for use where gas engines furnish the motive power, still more do they compete on equal terms as regards accuracy, whilst at a considerably lower cost, with those instruments of foreign manufacture which have up to date been so frequently employed where their extra cost as compared with the total cost of the installation is of no account.

Hot-wire instruments for direct and alternating currents.

Hot-wire Instruments are extremely suitable for use with both direct and alternating currents, are absolutely unaffected from any external source and, as compared with direct current electromagnetic instruments, are free from the residual magnetism error arising in the latter, and therefore an accuracy of at least $\frac{1}{2}$ per cent is attainable; as the prices for these instruments are but slightly above those for electromagnetic instruments they should be preferred to the latter. For alternating current they are superior to all others as they are free from self induction and entirely independant of the number of cycles. The measuring wire in these instruments is not damaged by currents twice as large as the maximum marked on the scale.

For installations in which large current variations occur it would therefore appear more adviseable to employ, even for alternating current, instruments of the **electromagnetic** type as they would not be damaged if the current for a short space of time very considerably exceeded the maximum range of the instrument. The type of electromagnetic instrument described on page 103 is better adapted for alternating current work than any other electromagnetic type now used, as the calibration curves for both classes of current are approximately alike and the number of alternations has practically no effect on the readings.*) It is however always preferable when ordering instruments in which an iron core is employed, for use with alternating current, to state the number of cycles.

Electromagnetic instruments for alternating current.

For electrical installations on board ships the use of the dead-beat standard voltmeter page 95 or the hot-wire voltmeter page 99 is preferable to that of an electromagnetic voltmeter with gimbal suspension usually adopted, as with the two former this inconvenient method of suspension is unnecessary, whilst the readings remain accurate owing to the aperiodicity of these two classes of instruments.

Marine instruments.

The **electrostatic** voltmeter for high potentials, page 109, the **Ohmmeter,** page 117, and the **Wattmeter,** page 118, should be specially noted; externally they are identical with the other instruments usually mounted on switchboards, are direct reading, and suitable for use with either direct or alternating current of any periodicity.

Electrostatic Voltmeter, Ohmmeter, Wattmeter.

All our instruments are carefully tested before they leave our works, and are only locked under lead seal with lettering H. & B. when they correspond with the readings of the check instruments which have been certified by the Imperial Physico-Technical Laboratory. Only in cases where this seal is not damaged and where there is no evidence of damage arising from unskilled use we undertake repairs free of charge.

In reference to our guarantee please refer to paragraph No. 9 on page V of the preface.

Guarantee.

If desired, these instruments will be supplied stamped and certified as to accuracy by the Imperial Physico-Technical Laboratory and the original documents delivered with them upon payment of the charges incurred (from 5 to 10 M.)

Certificate.

---

*) On the use of electromagnetic measuring instruments for alternating current see Dr. Bruger, Bericht über die Verhandlungen des Int. Elektrotechniker-Congresses zu Frankfurt a. M. 1891. Sekt. 1, Seite 89.

No. 601. Standard dead-beat current and potential indicators for direct current,

for use as portable control instruments.

Arrangement
of details patented.

Size:
18×19×10 cm.

These **Volt-** and **Ampèremeters** are based, like the Weston instruments, on the principle of the Deprez-d'Arsonval galvanometer, resembling therefore those described on page 14 and page 28 and therefore consist of a moveable coil, in this case pivotted in jewels, working in a very powerful homogeneous magnetic field. The moving parts are accurately balanced and the instrument can therefore be used in any position either vertical or horizontal. The special features of this class of instrument are the proportional scale, extreme sensitiveness throughout the entire range, dead-beat movement of the pointer and freedom from errors due to external currents; for use as portable instruments they are mounted in strong oak case with handle.

No. **601. Combined standard dead-beat current and potential indicators** for currents up to 30 ampères max. and for potentials up to 200 volts max. with any desired range, 100–150 graduations on scale . . . . **M. 225.—**

Commutator for above for two degrees of sensitiveness in potential measurements (e. g. up to 15 in tenths of a volt and to 150 in single volts) . . **M. 25.—**

No. **601a. Dead-beat control ampèremeter** up to 30 ampère max.
    I. with any desired degree of sensitiveness . . . . **M. 150.—**
    II. with two degrees of sensitiveness (e. g. up to 1.5 in one hundredths of an ampère and up to 30 in fifths of an ampère) . . . . **M. 180.—**

No. **601b. Dead-beat control voltmeter** up to 500 volts max.
    I. with any desired degree of sensitiveness . . . . **M. 130.—**
    II. with two degrees of sensitiveness (e. g. up to 100 in single volts and to 500 by five volts) . . . . **M. 150.—**
    III. with three degrees of sensitiveness . . . . **M. 175.—**

Shunts for use externally with Nos. 601 and 601a up to 150 ampères and **Additional resistances** for Nos. 601 and 601b up to 1500 volts . **M. 40—100.—**
Leather case with straps . . . . **M. 20.—**

*) In ampèremeters with one degree of sensitiveness and in all the voltmeters the terminals are now mounted inside the case, they are therefore inaccessible when the lid is closed.

No. 602 and 603. Standard dead-beat
Ammeter and Voltmeter for direct current,
in circular case for switchboards.

$^1/_4$ n. Gr.

Patented
arrangement of
details.

Diameter
of brass base
225 mm.

The system of a moveable coil in a magnetic field employed in No. 601 is here mounted in a well got up circular brass case. This system has the great advantage over all electromagnetic systems in which iron is employed to measure large currents, that it is unaffected by any currents in its proximity and the results of residual magnetism are entirely eliminated and as the pointer takes its position instantaneously also renders it suitable in cases where other instruments owing to the pulsating character of the current cannot be employed. The use of instruments of this class is always preferable where absolutely accurate measurements are required.

Standard dead-beat Ammeters, circular type
with exactly equal scale divisions from zero.

The terminals are at top and bottom of the case as in other ammeters.

No. **602**	I	II	III	IV	V	VI	VII	VIII	For larger currents special quotation
For amp. max. .	0,5 or 1	2, 3 or 5	10; 20, 30; 50 or 75	100 or 150	200 or 250	300	400	500	
Divided in amps.	0,01	0,02 and 0,05	0,1; 0,2; half	single	from 2 to 2		from 5 to 5		
Price in M. . .	**110**	**115**	**120**	**140**	**150**	**160**	**170**	**180**	

No. **602a. Similar ammeters** but with the **zero in the centre** of the scale, indicate therefore the direction of the current, with half the number of graduations for the above maxima of ranges. Extra M. **5.—**

Standard dead-beat Voltmeters, circular type

with very open scale at the required reading and omitting the lower values. | with exactly equal scale divisions from zero.

No. **603**	I	IA	II	IIA	III	IV	V	VI	VII	For higher readings special quotations
Up to volt . .	50—75	60—90	90—120	100—160	1; 3 or 5	10; 20 &c. to 50	100 or 200	300 or 400	500 or 600	
Divided in volts	half or single volts		single		0,01;0,05	0,1; 0,2 or 0,5	$^1/_1$ and $^2/_1$	from 2 to 2	5 to 5	
Price in M. .	**115**	**115**	**120**	**120**	**110**	**115**	**120**	**130**	**140**	

No. **603a. Similar voltmeters** but with the **zero in the centre** of the scale, deflecting therefore in both directions, with half the number of graduations. Extra M. **5.—**

Commutator to increase the range of the voltmeter for lower tensions by dividing the additional resistance placed inside the instrument. Extra M. **30.—**

No. 604.
Standard dead-beat instruments for small currents and low potentials
in upright cases.
Arrangement of details patented.

The magnets and the moveable coil, pivotted in jewels, are mounted in a polished mahogany case. This design is specially adapted for medical requirements, though it is frequently employed in telegraphic work to determine the insulation of the line and the comparative strengths of the flow and return currents, for which purposes it is more convenient than the differential galvanometer usually employed.

No. **604. Standard dead-beat instrument,** in upright case for small currents, with any desired range up to say 1000 milliampères in 20—25 equal divisions on both sides of the zero, or for low potentials with any desired range not exceeding 25 volts. Price M. **95.**—

No. **604a. Instrument as above** with either lever- or plug switch and two ranges for current and potential measurements (e. g. to 20 and 200 milliampères or to 2,5 and 25 volts). Price M. **125.**—

No. **604b. Instrument as above** with either lever- or plug switch and three ranges for current and potential measurements (e. g. to 20 milliampères, 200 milliampères and to 100 volts). Price M. **140.**—

No. 605a.
Dead-beat accumulator cell tester.

Arrangement
of details patented

Diameter of case
80 mm,
depth 40 mm.

³/₄ n. Gr.

No. 605a I.

The moveable coil magnetic system described in No. 601 is here mounted in a small circular case; the instrument can be connected to any desired cell regardless of the direction of the current as it deflects to both sides and can be read in any position.

No. 605.	I. to 3 volts in tenths	II. to 10 volts in half volts	III. to 25 volts in single volts
Price in M.	56	58	60

No. 605a. Small Standard Milliampèremeter.*)

This instrument, which in every respect resembles No. 605, has several very considerable advantages as compared with those usually employed in electropathy: 1) no adjustment to the magnetic meridian is necessary, 2) it is unaffected by other currents in the neighbourhood, 3) dead-beat pointing, it therefore gives instantaneously the correct reading without vibration even with the maximum variation in the quantity of current, 4) equally divided scale and equal degree of sensitiveness throughout the entire range, 5) is also available as voltmeter for testing single cells or accumulators, 6) low internal resistance.

No. 605a.	I. to 25 milliamp.	II. to 35 milliamp.	III. to 35 and with shunt to 350 M.A.
Price in M.	60	64	83

No. 461a. Large Standard medical Milliampèremeter.*)

The moveable coil magnetic system described in No. 601 is here mounted in a large polished mahogany case, see illustration page 105, and has all the advantages described in No. 605a; it is fitted with a long pointer moving over a scale of 16 cm radius, and the graduations can be read at a distance of several metres, thus enabling the measurements arising from the application of the electrodes to be easily checked. It is graduated to read direct from 1 to 20 milliampères and by means of a shunt the range can be extended either 10 or 20fold as ordered, that is to 200 or 400 milliampères. The zero is in the centre of the scale and the instrument can therefore be used irrespective of the direction of the current. Price M. **150.**—

*) For other medical galvanometers see pages 100 and 105.

No. 600. Hot-wire Ammeters and Voltmeters for direct and alternating current,
for use as portable control instruments.

¼ n. Gr.

System
patented.

Size:
20 × 19 × 11 cm.

In these instruments a short wire of platinum silver alloy is employed to carry the current, and the elongation of this wire arising from the heating effect of the current is transmitted by means of a special arrangement to a pointer mounted in jewels. Means are provided outside the case to enable the pointer to be brought back to zero, should the necessity for so doing arise through overstraining the instrument, this does not affect the graduations. These instruments, compared with others working on the same principle, absorb but little current in the measuring wire and therefore the readings are permanently correct, this is confirmed by the results of a series of tests made by the Imperial Physico-Technical Laboratory. A magnetic damping arrangement ensures the dead-beat action of the pointer. These caloric instruments as compared with all electro-magnetic instruments have the advantage of being absolutely unaffected by external currents, and can be used for either direct or alternating current with the same scale.

No. 600. Dead-beat hot-wire Ammeters and Voltmeters combined, in polished oak cases for currents not exceeding 50 ampères and potentials up to 200 volts max. with any desired range, 50—80 divisions on scale.
M. **180.**—

If with **commutator** to increase the lower range of the scale for potential measurements; by dividing the resistance. Extra M. **30.**—

No. **600a. Dead-beat hot-wire Ammeter** up to 50 ampères max. as portable control instrument
I. with any desired range of scale . . . . M. **115.**—
II. with two ranges (e. g. to 3 and 30 amp.) . . . . „ **145.**—

No. **600b. Dead-beat hot-wire Voltmeter** up to 200 volts max.
I. with any desired range of scale . . . . M. **110.**—
II. with two ranges. . . . . „ **140.**—

Leather case with carrying straps . . . . „ **20.**—

No. 598 and 599. Hot-wire Ammeter and Voltmeter for direct and alternating current

in circular cases for switchboards.

System patented.

No. 598.   ¼ n. Gr.   No. 599.

¼ n Gr

Brass base 225 mm diameter.

On the same system as No. 600 but mounted in a circular brass case. As is well known the readings of most voltmeters and ammeters are affected by external magnetic fields, from which it is sometimes impossible to fix the instruments at a sufficient distance; and in such cases it is advisable to use these caloric instruments even in direct current installations as they are absolutely unaffected thereby. They are the most trustworthy instruments for alternating current work as their readings are independant of the current curve and the periodicity of the installation.

Dead-beat hot-wire Ammeter in circular case
with approximately equal divisions on scale.

No. 598.	I	II	III	IV	V	VI	VII	VIII	For larger currents special quotations.
For amp. max.	0.3, 0.5, or 1	2, 3 or 5	10, 20; 30, 50; or 75	100 or 150	200	300	400	500	
Divided in amp.	0.05 and 0.02	0.1 and 0.2	¹/₅; ¹/₁; ²/₁	⁵/₁ or ⁵/₁	fr. 5 to 5	from 10 to 10	fr. 20 to20		
Price in M. . .	75	80	85	85	90	100	110	120	

The lower limit of range is approximately 10% of the upper.

The fall of potential at the extreme limit varies in the different ammeters from 0.2 to 0.3 volt.

The measuring wire stands without damage double the normal maximum current marked on the instrument; should the zero of the instrument vary through such overloading the pointer can be reset by means of the adjusting screw provided for this purpose on the left hand side of instrument.

Dead-beat hot-wire Voltmeter in circular case
with open scale for ¹/₅ volt   with approximately equally divided scale
at the required reading.   except when otherwise ordered.

No. 599.	I	IA	II	IIA	III	IV	V	VI	VII	For higher readings, with transformers if desired special quotations.
For or up to volt	50-75	60-90	90-120	100-150	1, 3 or 5	10, 20; 30, 50	100 or 200	300 or 400	500	
Divided in volt	below the required reading from 5 to 5 and then in ²/₁				0.02 and 0.1	half; single	fr. 5 to 5	fr. 10 to 10	fr. 20 to 20	
Price in M. .	75	75	75	75	75	75	75	95	100	

The current employed at the extreme limit of the voltmeter is about 0.22 ampère.

Commutator to extend the range by dividing the additional resistance of the instrument.   Extra M. 30.—

No. 379 and 379a. Spring galvanometers
Kohlrausch type.

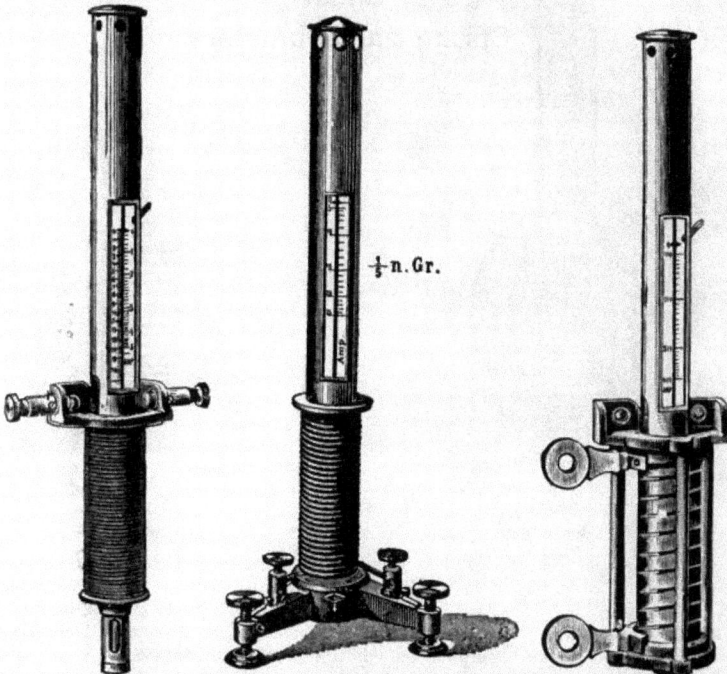

$\frac{1}{5}$ n. Gr.

No. **379** II.   No. **379a** III.   No. **379** VII.

A thin soft iron tubular core, suspended on a torsionless spiral spring is sucked into a solenoid. At the base is a projecting rod which acts as a guide to the core and at the same time exerts a powerful damping action without affecting the accuracy of the readings. The maximum movement of the core is approximately 100 mm; the left hand side of the scale being divided into millimetres the amount of movement of the core can be directly measured, whilst the right hand side is graduated to the equivalent in ampères. By means of various patented forms of iron core the divisions on the scale can be varied to suit requirements.

Spring galvanometer For amps. max.	I*)	II	III	IV	V	VI	VII	VIII	IX	X
	0,001 to 1	2, 3 or 5	10, 20 40 or 60	80, 100 or 120	200	300	400	500 or 600	700 or 800	1000
No. **379** Wall pattern M.	—	60	54	60	110	125	150	180	250	320
No. **379a** on tripod base M.	75	76	70	76	Instruments for larger currents can only be supplied in the wall pattern.					

The minimum current shown is approximately 10% of the maximum.

*) When this instrument is specially supplied for medical purposes a steel needle replaces the iron core and the scale is graduated to read direct in milliampères; by plugging in shunts this range can be increased 10 or 20 fold. For galvanometers for medical purposes see pages 97 and 105.

No. 380 and No. 380a.

Small
Spring galvanometers.

Patented form of iron core.

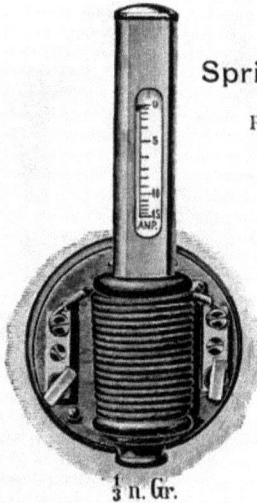

$\frac{1}{3}$ n. Gr.

$\frac{1}{3}$ n. Gr.

No. 380. | No. 380a.
To fix against wall. | On wood base.

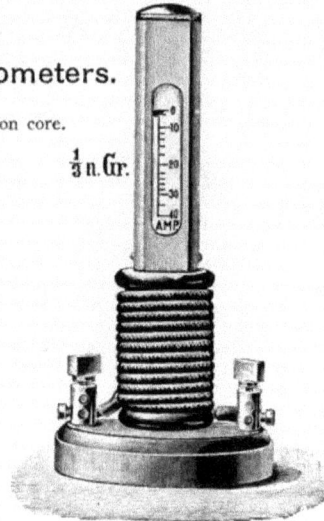

These instruments are fitted with patented form of iron core; they are simply and cheaply made, mounted on polished mahogany base and with nickelled scale tube, they are suitable for constant use in connection with subcircuits of an installation, as control instruments in arc light circuits, as lamp counters, and for laboratory work. In both designs at the same price and with the following graduations.

No. 380 and 380a	I	II	III	IV	V	VI	VII	VIII
For amp.	0.5—2	1—5	2—10	3—15	4—20	5—25	6—30	10—40
Divided in amp. . .	0.1	0.2	0.5	single			from 2 to 2	
Price in M. . . .	25.—	20.—	15.—	15.—	15.—	15.—	15.—	20.—

The above ammeters, graduated for alternating current . . . Extra M. 3.—

No. 522.
Current Index.

For use on switch boards in installations with accumulators, indicating charge and discharge with the smallest current; well finished and mounted in brass case

I. for currents up to 200 ampères . . . M. **12.**—
II. „ „ „ „ 500 „ . . . „ **20.**—

$^{1}/_{4}$ n. Gr.

Electromagnetic Ammeters.

Patented form of iron core.

No. 379b.

No. 379c.

¹/₄ n. Gr.

Diameter of brass base 155 mm in the small
and 225 mm in the larger sizes.

The design of these **Ammeters** for **direct current** is the same as that of
the Kohlrausch spring galvanometer, namely an iron core sucked into a solenoid, but
in this case a spring in torsion acts as the opposing force. The straight line motion
is converted into a circular one by a lever acting on an axle pivotted in jewels.
A **patented form of iron core** enables a scale with equal divisions to be attained.
These instruments resemble externally the voltmeters described on the following page
and they are of excellent mechanical design and first class finish.

No. 379b and 379c. Ammeters for direct current
with approximately equally divided scale.

No. 379.	I	II	III	IV	V	VI	VII	VIII	IX	X	XI
For amp. max.	0.5 or 1	2 or 5	10, 20, 30 50 or 75	100 or 150	200 or 250	300	400	500 or 600	700 or 800	1000	2000
Divided in amp.	0.02 and 0.05	0.1 and 0.2	¹/₂, ¹/₁ and ⁶/₁	⁵/₁ or ⁵/₁	from 5 to 5	from 10 to 10	from 20 to 20		from 50 to 50		
Price b small size	65	55	45	50	55	65	are not made in the small size				
in M. c larger size	70	60	50	55	60	70	75	90	130	175	250

The lower limit of range marked is as a rule 10% of the maximum value.

In many cases instruments with a longer vibration period are preferable, such
as ammeters constructed on the same principle as the voltmeters described on the next
page: This type is suitable for both **alternating** and **direct current** and is made in
both sizes above described as

No. 379d and e. Ammeters for direct and alternating current
in small (d) and larger (e) sizes with the same scales as above.

No. 379.	I	II	III	IV	V	VI	VII	VIII	IX	X	XI
Price d small size	70	60	50	55	60	70	are not made in the small size.				
in M. e larger size	75	65	55	60	65	75	80	110	150	200	280

Ammeters for **larger currents** specially quoted for.

Ammeters for **high potential** work exceeding 1000 volts can be supplied with special insulation arrangements
at an extra cost; we make in this case the leads in the form of bolts projecting back through the switchboard so that no
parts carrying current are accessible.

Electromagnetic Voltmeters.

No. 381b. System patented. No. 381c.

$^1/_4$ n. Gr.

Diameter of the brass base 155 mm in the small
and 225 mm in the larger sizes.

By means of the combined effect of two or more extremely light tubular segments of soft iron acting on each other it is possible to obtain a scale with very open graduations at the usual working position or one with approximately equal divisions throughout the entire range. The hysterisis error is extremely small owing to the shortness of the magnetic axes in the iron cores. The arbor is in aluminium with steel points working in jewels. Externally the instruments resemble the ammeters described on the previous page and the instruments are mounted in well made and highly finished cases. These voltmeters can be used with the above scales for either **direct** or **alternating current** of any periodicity, but are always calibrated without extra charge with whichever class of current is desired.

No. **381b** and No. **381c**. **Voltmeters** for **direct current,**
No. **381d** and No. **381e**. **The same** for **alternating current,**
with very open scale for $^1/_1$ Volt at the usual working positon.

No. **381.**		I	I A	II	II A
For volts		50 to 75	50 to 90 or 100	90 to 120 or 130	100—150 or 160
Price in M.	**b** or **d** direct current, small size	53	53	55	55
	c or **e** alternating current larger size	58	58	60	60

The same but with approximately equally divided scale, unless otherwise ordered.

No. **381.**		III	IV	V	VI	VII	VIII	IX*)	X	IX	XII
To volts max.		3 or 5	10,20 or 30	100 or 200	300 or 400	500 or 600	700	800	1000	2000	3000
Divided in volt		0.1	$^1/_2$ and $^1/_1$	from 2 to 2	from 5 to 5	from 10 to 10			from 20 to 20		fr. 50 to 50
Price in M.	**b** or **d** dir. current small size	60	60	65	are not made in the small size						
	c or **e** alternating current larger size	65	65	70	80	90	100	125	145	175	225

*) The additional resistances for instruments from 800 volts upwards are contained in a square metal case 25 × 25 × 6 cm upon which the instrument is mounted.

Red Pointer for indicating the normal voltage, greatly facilitating reading at a distance, for the large size only . . . . M. 3.—

Extension of range by switching part of the winding in series or parallel, with commutator and two scales
to **double** the range. e. g. the first to 90 and the second to 180 volts {only for the M. 30.—
to **quadruple** the range. e. g. the first fr. 5 to 50 and the second to 200 volts{ larger size } „ 40.—

No. 381. Portable Voltmeter.

System patented.

$^1/_3$ n. Gr.

The same magnetic system as described on the previous page is here mounted in a brass case which is fitted with three feet, one of which is arranged as levelling screw to adjust the pointer to the zero position.

No. 381. Portable Voltmeter

No. 381	with open scale at working position.				with approximately equal divisions.				
	I	IA	II	IIA	III	IV	V	VI	VII
Range in volts .	50—75	50—90	90—130	100—150	0.5—5	5—25	10—50	20—100	50—200
Price in M. . . .	70	70	70	70	75	75	75	75	80

No. 381 M. **Inspector's voltmeter,** similar in design to above, with two scales; at the back of the instrument is a commutator to place the resistance corresponding with either the one or the other scale in series, one scale is to 90 and the other to 150 or 160 volts . . . . M. **90.—**

Lock up **travelling case** for the above . . . . „ **12.—**

Commutator to extend the range of the other portable voltmeters No. 381 III—VII, with two scales, e. g. 0.5—5 and to 25, or 10—50 and 50—200 volts. M. **25.—**

Marine-Voltmeter.

No. **381 a D. Ship voltmeter,** similar principle to the above but with gimbal suspension and ceiling plate, in two ranges, I for 65 volts, II for about 100 volts. Price M. **85.—**

No. **381 a W.** As above but with gimbal suspension on a bracket. M. **85.—**

Ammeters cannot be supplied to hang in gimbals, nor are they requisite for voltmeters if those types of **Hot-wire instruments** described on page 99 or the **Standard instruments** on page 95 are selected, as both the ammeters and voltmeters of these types give correct readings in any position and are therefore unaffected by the ship rolling or pitching.

$^1/_3$ n. Gr.

No. **381 a D.**

Simple Volt- and Ampèremeters

for low tensions and small currents.

System patented.

No. 382.

1/5 n. Gr.

No. 460a.

In these **simple Voltmeters and Ammeters** the same electromagnetic principle is employed as in those described on the two previous pages, and the most important parts are of the same high class of workmanship; the arbors for instance being also mounted in jewels. The scales however are not so closely divided and are so arranged as to be legible at a considerable distance. The cases are of highly polished mahogany.

No. **382.** **Simple Voltmeters** for low tensions, with backboard, specially suitable for nickelplating and galvanoplastic work, in the following ranges all at the same price.

I. 0.5—3 volts divided in ²/₁₀; III. 2—10 volts divided in ¹/₁ |
II. 1—5 „ „ „ ⁵/₁₀; IV. 4—20 „ „ „ ²/₁ | . Price M. **25.**—

No. **382a.** **The same instruments** in portable case with handle; suitable for cell and accumulator testing, specially for testing electric bell-, clock-, or telephone installations, &c., with the same ranges as No. 382. Price M. **30.**—

No. **460.** **Simple Milliampèremeters** in portable case with handle, very suitable for **medical purposes.***) In the following ranges all at the same price.
I. 2—12 m.-amp.; II. 4—20 m.-amp.; III. 6—36 m.-amp. M. **35.**—

No. **460a.** **The same instruments** with **plug switch** to add shunts, thereby increasing the range tenfold, that is to 120, 200, and 360 milliampères. M. **45.**—

No. 461. Large Milliampèremeter.

Large Milliampèremeter*) in flat polished mahogany case, with scale of 16 cm radius and graduations easily legible at a distance of several metres. Direct reading up to 20 milliampères, and with shunt in circuit to either 200 or 400 milliampères: direction of current does not affect the readings.
Price M. **150.**—

1/6 n. Gr.

*) For **Standard** instruments for medical purposes see page 97.

No. 383.
Voltmeter for direct or alternating current,
large size,
for main switchboards in central stations.

Diameter
of the frame board
40 cm.

Diameter
of the case
35 cm.

$^1/_6$ n. Gr.

For large electrical installations, especially where a network of mains is fed from several different sources, the use of a main potential indicator not resembling the other measuring instruments is advisable. In all large installations also it is desirable that the engineer in charge should be able to check the potential maintained in the mains, which is his chief duty to keep constant, from any part of the engine room. This voltmeter complies with these requirements.

No. **383. Voltmeter, large size,** in brass case framed in a polished wood base. Length of pointer 17 cm; the scale is only graduated for + 20 volts of the normal potential so that the width of each division for 1 volt is about 1 cm and the instrument can therefore be easily read at a distance of 6 to 10 metres. The system employed for **direct current** instruments is that of a moveable coil in a magnetic field (see page 94) whilst that used for **alternating current** works on the caloric principle (see page 98), our improvements in both systems are patented.

The price is the same for both classes of current and the following potentials:

I. to 200 volts M. **200**; II. to 500 volts M. **225**; III. to 1000 volts M. **250**.

Ammeters mounted in similar large cases specially quoted for.

No. 383a.
Universal Voltmeter for direct or alternating current
suitable for Central Stations.

¹⁄₆ n. Gr.

No 383a. **Voltmeter,** l a r g e s t s i z e, in brass case 50 cm diameter framed in polished walnut base of 53 cm diameter. As in voltmeter No. 383 the entire range of the scale is graduated in 30—40 divisions, therefore for 15—20 volts over and under the normal potential stated on order; each division for 1 volt is therefore about $1^1/_2$ cm wide and the divisions are so heavily marked that the instruments can be read accurately from a distance of 10—15 metres. The pointer is 25 cm long and is so close to the scale that, even if the instrument is mounted at a considerable height, the parallax error is eliminated. The **direct current** instruments are made on the principle of the standard voltmeter, and those for **alternating currents** on the hot-wire principle. In both systems the freedom from hysterisis, and from external magnetic influences and their aperiodicity are special features.

The price is the same for both classes of current and the following potentials:

I. to 200 volts M. **300.** — ; II. to 500 volts M. **325.** — ; III. to 1000 volts M. **350.** — .

If sufficient time is allowed the large voltmeters No. 383 and No. 383a can be supplied mounted on any other sort of wood or on metal.

107

No. 534. Accumulator cell tester.
Pocket voltmeter.

System patented.

²/₃ n. Gr.

This pocket instrument is constructed on the principle of a narrow solenoid acting on a pivotted ring-shaped curved magnet, and resembling therefore that employed in those described on pages 29 and 118. Special care is taken in making and testing the constancy of the magnet employed. The small voltmeter is very suitable for testing separate cells in a battery of accumulators and can be used in any position. Owing to its comparatively high resistance it can be kept in circuit for some time; it is mounted in ebonite case with thick glass in both sides and fitted in a pocket case.

I. For potentials up to 3 volts . . . . Price M. **36.—**
II. For any desired potential up to 15 volts max. . „ „ **44.—**

For permanent use in accumulator installations a stout pointed copper wire is clamped under one terminal and contact made thereby with the corresponding pole of the accumulator; a flexible cord is connected with the other terminal and contact therewith to the other pole of the cell. To avoid damaging the instrument I by excessive current it must never be connected to more than one accumulator cell.

No. 503. Hydrometer for Accumulators.

Hydrometer with flat container to determine the density of the acid in accumulators, can be easily inserted between the plates in a cell and has a legible scale gratuated on one side to Beaumé and on the other to specific gravities.   Price per pair M. **3.75**

No. 505. Magnetic field explorer.

½ n.Gr.

A magnetized needle is pivotted in strong framework so that it is free to move in any direction and provided with a telescopic handle; this is a very practical instrument for testing dynamos and as a pole finder is more convenient than pole testing paper as the direction of current in insulated wires can thereby be ascertained in accordance with Ampère's rule. With leather case.
Price M. **10.—**

No. 594.
Electrostatic Voltmeter for high potentials.

Brass base   225 cm diameter.

$\frac{1}{4}$ n. Gr.

These static voltmeters for high potentials are designed on the principle of the Thomson quadrant electrometer, externally they resemble the other types of instruments in circular cases. A biscuit shaped needle moves between four pairs of quadrant shaped inductors which are specially designed to secure an approximately proportional scale. Two diametrically opposed pairs repel the needle and the other two attract. The voltmeters are suitable for measurements of direct and single or multiphase alternating currents of any frequency and also for static electricity, absorb no power, are free from self-induction and unaffected by external currents. An aluminium disc working between the poles of a permanent magnet is mounted on the arbor of the pointer to render it dead-beat. Easily fusible wires are fixed in the leads to the quadrants and the needle, and a spark gap is provided between them so that in the event of an abnormal rise of tension a spark passes across and protects the instrument from damage. All parts carrying current are extremely well insulated from the case.

Electrostatic **Voltmeter** for high potentials.

No. 594.	I	II	III	IV	V	Higher readings
For volt max. .	1000	2000	3000	4000	5000	to 10000 volts specially
Price in M. . .	120.—	150.—	175.—	200.—	225.—	quoted for.

The lowest reading is approximately 10% of the maximum range of the instruments. Instruments for intermediate ranges are charged as for the next higher range.

Contact Voltmeter for direct and alternating current
with patented form of iron core.

No. 510.

Contact Voltmeter $\frac{1}{4}$

mounted on metal base aat.Gr

in lock up cover

of polished walnut.

Prices below.

Size

of the metal base

200×155 mm.

It intended for alternating current,
kindly state number of cycles.

These contact voltmeters are made with frictionless iron core and air damping arrangement, a very great movement at the usual working pressure being attained by the patented form of iron core employed. On this core platinum contact pieces are mounted which if the pressure varies make good contact in either direction, thereby closing the local circuit. The apparatus is especially intended for working optical or accoustic signals, and for controlling circuits in which automatic pressure regulators or cell switches are placed, the relay No. 511 on the following page being sometimes used in connection therewith.

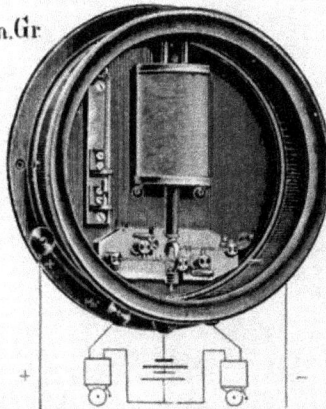

$\frac{1}{4}$ n. Gr

No. 510 a.

Contact Volt-

meter

in circular case,

matching externally

our volt- and ampère-

meters.

Diameter of the brass

base 225 mm.

The front of case is

easily removeable, to

allow easy adjustment

of the platinum tipped

spring contacts.

Calibrated with either direct or alternating current	I	II	III	IV	For higher
For volts	50—75	100—130	250	500	pressures
Price in M. No. 510	65.—	65.—	—	—	special
No. 510 a . . .	65.—	65.—	80.—	100.—	quotations.

A sensitiveness of $1-2^o/o$ is necessary to ensure good contact, a variation of $^1/_2-1^o/o$ over or under the usual normal pressure can therefore be indicated. The maximum current admissible through the contacts is $^1/_2$ amp. For optical signals lamps taking small current must be used or, if larger currents are necessary, a relay (see following page) must be inserted in the circuit.

Relay for direct and alternating current.

No. 511.

Size of base
120 × 80 mm

½ n. Gr.

No. 511. **Single relay** tor contact voltmeter with adjustment for armature.
For 65 or 100 volts . . . . Price M. 30.—

By means of this relay which owing to its high resistance absorbs but very little current, currents up to 2 ampères can be employed in the local circuits controlled by the contact voltmeter.

The two terminals in front to which the coil windings are joined are connected to the contact voltmeter or source of current, the two at sides are connected to the circuit in which the large current is required.

¼ n. Gr.

No. 511a.

can be used in either a horizontal or vertical position.

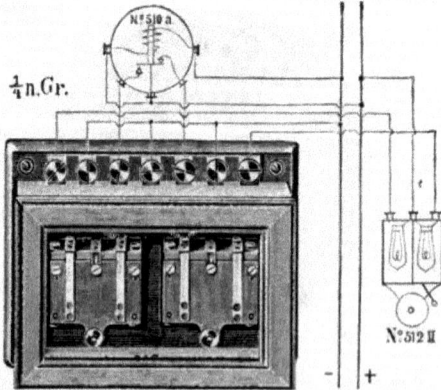

Size of back
200 × 155 mm

N° 512 II

No. 511a. **Compound Relay** (double) tor contact voltmeter, similarly constructed to the above, mounted on metal base in glass fronted polished walnut case, for use with No. **510** (see previous page). For 65 or 100 volts.   Price M. **70.—**

No. 511b. As above, but in circular case for use with No. **510a.** Price M. **80.—**

If two different local circuits are to be controlled by the contact voltmeter, or if the direction of current in the local circuit is to be reversed, when the pressure is too low, from that employed when the pressure is too high, then two relays are necessary. The above illustration shows the method of connecting a compound relay with a contact voltmeter and an optical and sound alarm signal No. **512 II** (see following page).

No. 512.
Signal apparatus with optical and sound alarms
to notify pressure variations and earth contacts.

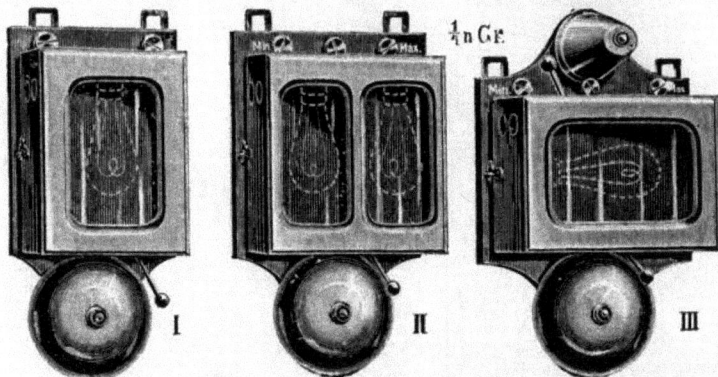

No. 512. **Signal apparatus with optical and sound alarms,** comprising an electric bell of low resistance in series with incandescent lamps, the latter mounted in polished walnut wood cases behind coloured glass. The apparatus is chiefly intended for use with the contact volt-meters illustrated on page 110, to which it can be connected without a relay provided that incandescent lamps taking less than $1/2$ ampère are used.

Made in three patterns, each for 65 or 100 volts (voltage to be stated on order).

I. with one bell and one lamp . . . . Price M. **25.—**

II. with one bell and two lamps, one of which is mounted behind green and one behind red glass, and in accordance with the variation of pressure the corresponding lamp is lit, the bell in either case continuing to ring . . . . Price M. **33.—**

III. with two bells of different tones and one lamp, one bell sounding at the minimum the other at the maximum pressure permissible whilst the lamp is lit behind red glass in either case.
Price M. **43.—**

The incandescent lamp holders are for screw socket lamps.

Earth contact alarm apparatus.

The **Signal apparatus No. 512 I** can also be employed as an **earth leakage indicator** if one terminal is connected to earth and the other either permanently connected to one pole or alternately to the + and — pole. The lamp begins to glow and the bell to ring if the insulation resistance falls below 1000 ohms.

No. 513. Alarm Voltmeter

with patented form of iron core.

¹/₄ n. Gr.

Case in polished
walnut wood.

Size
of backboard
36×24 cm.

No. **513. Alarm Voltmeter.** In this instrument the motion of an iron core in a solenoid through which the current is passed, is transmitted by a lever to an arbor fitted with a spiral torsion spring; the core also is suspended on a long spiral spring. Owing to the core being mounted between two springs the arbor does not carry its weight and therefore the index pointer assumes the accurate position at any part of the scale. A second lever is mounted on the arbor which touches one or other of the two contact screws fixed on the right and left hand, thereby closing the circuit of a relay which works the alarms; these latter are a bell and lamps which give red or green light according to whether the maximum or minimum pressure permissible has been exceeded. The contact point for minimum pressure is arranged to be turned aside from outside the case so that the pointer may, when no current is passing through the solenoid, assume its zero position, both contact screws can also be adjusted for height so that the alarm circuit may be closed at any desired pressure.

For 65 or 100 volts including bell*) and incandescent lamps.**)   Price M. **175.—**

For 200***) volts . . . . „   „   **185.—**

If the bell is not required, deduct . . . . „   „   **12.—**

*) In most cases it is preferable not to mount the bell on the apparatus but to fix same at the most suitable spot.
) The lamp sockets are for Edison screw fittings, and the lamps are mounted behind coloured glasses; ordinary clear glass lamps of **5 c. p. (to avoid heating the apparatus) should therefore be used.
***) In this case, as **200** volt lamps are not always in stock, another **100** volt lamp is placed in the alarm circuit.

Differential Voltmeter for direct and alternating current.

¼ n. Gr.

No. 606.

Diameter
of the brass base
155 mm.

Differential voltmeters are used to indicate the difference in pressure between two sources of current in the three wire system or in two machines to be connected in parallel, also to show the loss in pressure in the various branches of a system of mains.

No. 606. The **differential Voltmeter** for direct current is made with a solenoid having two separate windings of equal number of turns and equal resistance through which the current flows in opposite directions. If the action of one coil is more powerful than the other it attracts the ⌃ shaped magnet which is mounted on an arbor within the solenoid. The scale can be divided into single volts.

For tensions of about 65 or 100 volts . . . . Price M. **105.—**

No. 606a. **Caloric differential Voltmeter** for direct and alternating current. This instrument is made with two hot-wire voltmeters (see pages 98 and 99) in which the pointer axes are in the same horizontal plane and whose pointers work in opposite directions over the scale. The scale is divided from the centre outwards with very open readings at the working pressure and is so proportioned that the divisions corresponding thereto are in the same radius and the pointers therefore at the usual working pressure overlap. With the differential voltmeters Nos. 606 and 514 only the difference of pressures can be determined — these instruments indicate direct how much the pressure in both branches is above or below the normal. This instrument is also much more suitable for other purposes than those which are already calibrated for some definite standard pressures.

For the usual voltages . . . . Price M. **125.—**

No. 5l4.
Differential-Voltmeter with alarm.

No. **514.** This **differential Voltmeter** with alarm is used, like the one described on the previous page to measure the drop of potential in leads, to check the difference in pressure on the two sides of a three wire system, for use when switching dynamos in parallel and for other purposes.

In this instrument two solenoids and not one as in the signal voltmeter No. 513 act on a soft iron core suspended between two springs the motion of the core being transferred by a lever on to the pointer arbor in a similar manner to that used in the other instrument. The two solenoids are wound with the same number of turns and also of the same resistance and are mounted so that their vertical axes are in the same straight line. The core projects equally into each of the solenoids and is so magnetised that in one a north pole and in the other a south pole of equal magnetic force is produced. On the least difference occurring in the magnetic effect of either solenoid the core is sucked further into either the one or the other, and the pointer connected to the arbor indicates the difference of potential existing at the terminals of the solenoids.

The contacts by which either one or the other of the two bells is actuated through a relay can be adjusted to work at any desired graduation on the scale. Price M. **175.**—

No. 608.
Back charge and current Index for working dynamos in parallel.

New system

Diameter of brass
base 225 mm.

¼ n. Gr.

When working dynamos in parallel, each driven by their own engine it sometimes occurs as is well known, that one dynamo drives the other as a motor, if the speed of the engine driving the latter falls below the normal. In working dynamos in parallel the poles also are sometimes reversed, which, when the machines are next started causes current to be delivered to the distributing mains in the wrong direction. In both these cases the faulty working of the machine is shown on the back-charge indicator by a red disc appearing in the centre of a white field. The apparatus works perfectly at about 5% of the maximum current. The design of this instrument resembles that of the ammeter page 102, but to answer both purposes above mentioned the solenoid is wound with fine wire for potential measurement as well as the usual winding for the maximum current, and in lieu of the soft iron core a permanent magnet is fitted which is sucked into the solenoid when the current is flowing in the right direction but repelled if the machines are working incorrectly. The general appearance externally is the same as that of the other instruments in circular cases for switchboard use.

Back charge Indicators for working pressures up to 130 volts

No. **608.**	III	IV	V	VI	VII	VIII	for larger
For amp. max.	30, 50, or 75	100 or 150	200 or 250	300	400	500	currents special
Price in M. .	80	85	90	110	120	130	quotations.

No. 612.

Leakage Indicator or Insulation tester for direct and alternating current

in circular case for switchboard.

$\frac{1}{4}$ n. Gr.

System patented.

Diameter of brass
base 225 mm.

Earth

+ −

Practically as is well known it is extremely difficult to avoid leakage in any large system of mains, but the loss of current arising herefrom should not exceed a definite amount. To avoid interference with the usual working routine the insulation resistance of the mains should be constantly checked which can be done without difficulty by this apparatus. The leakage indicator corresponds externally with the other switchboard instruments and shows the insulation to earth direct in ohms. One terminal of the instrument is permanently connected to earth (gas or water main) and the + and − poles of the dynamo or mains are connected to the other two terminals and either can be joined through the instrument to earth by means of the lever at side of case. If any considerable leakage is present, the instrument can be used to localise the fault.

Leakage Indicators for direct and alternating current.

No. 612.	I	II	III	IV	V	VI	VII	VIII	IX
Approximate tension in volts	65—75	100—125	200	300—400	500—600	700—800	1000	2000	3000
Range in ohms	60000 to 200	100000 to 300	200000 to 500	400000 to 2000	600000 to 10000	800000 to 20000	1 megohm to 30000	3 megohms to150000	4 megohms to300000
Price in M.	70.—	70.—	70.—	90.—	95.—	100.—	110.—	120.—	150.—

The readings of the pointer are only correct at a certain working pressure, which must therefore be stated on the order, but slight variations have only an inappreciable effect on the result.

If both poles are earthed the readings shown are somewhat too high, and if is desired to accurately determine the amount of leakage the following formula must be used. Let $\alpha +$ = the reading of the instrument when the lever is on + and $\alpha -$ when the lever is on − then the leakage to earth on the + pole is W_+ and that on the − pole W_-, when W represents the resistance of the instrument

$$W_+ = \alpha_- - \frac{w(w + \alpha_-)}{w + \alpha_+} \quad \text{and} \quad W_- = \alpha_+ - \frac{w(w + \alpha_+)}{w + \alpha_-}$$

No. 610. Direct reading Wattmeter for direct and alternating current

in circular case for switchboard.

¼ n. Gr.

System patented.

Diameter of brass base 225 mm

This instrument is constructed on the same electrodynamic principle as Nos. 531 to 533 page 29 without using iron, and reading direct on an approximately equally divided scale. The silk fibre suspension there employed is replaced in this instrument by a horizontal arbor pivotted in jewels and a very efficient air damping arrangement which renders the pointer practically dead-beat is substituted for the magnetic damper. The moving part is provided with an astatization coil as well as the working coil so that external fields but slightly affect the position of the pointer. As counter force is employed either a spiral spring or according to our new patent a quartz fibre. It is especially suitable to determine the watts absorbed by incandescent lamps whilst taking photometric measurements and for alternating current installations where the direct reading and quickness in action are great advantages. The instrument externally resembles our other circular cased instruments but the terminals for the main current are at the sides of the case and not vertically arranged.

Wattmeters arranged for a maximum potential*) of 100 volts.

No. 610	I	II	III	IV	V	VI**)	VII	VIII	IX
F. amp. max.	5	10	25	50	75	100	150	200	250
Kilowatt	0.5	1	2.5	5	7.5	10	15	20	25
Price in M.	170.—	170.—	170.—	180.—	190.—	250.—	270.—	290.—	310.—

No. **610a. As above,** but arranged for a maximum potential of 150 volts (i. e. for tensions exceeding 100 volts) . . . Extra M. **10.—**

Resistances in separate cases for use in series with No. 610, with all accessible parts carrying current cased in insulating material, for maxima potentials
of 500 volts M. **75.—**; 1000 volts M. **100.—**; 2000 volts M. **200.—**.

Series resistances for higher potentials and **Shunt resistances** for currents exceeding 250 ampères specially quoted for.

*) Kindly state working pressure and maximum current or quote roman number **when ordering.**
**) Shunts for currents of and exceeding 100 ampères are mounted in a special case.

No. 543.

Electrodynamic spring balance for the zero method.
(Wattmeter.)

The moving coil of the Electrodynamo-meter is mounted to turn easily within the main solenoid, a long but rigid aluminium pointer with its point working over a marked mirror renders it easily adjusted to zero; two stops, between which a lever connected to the moving coil plays, prevent the pointer moving too far from zero in either direction, on the end of this lever a horizontal disc is mounted which dips into an easily removeable cup filled with fluid, thereby causing a very efficient damping action. The movement of the coil when deflected by the current is opposed by a spiral spring in which the extension is directly proportional to the intensity of the current. The wire carrying this spring is, as in other known torsion instruments, wound on a cylinder, thereby extending the spring until the pointer again attains the zero position. The reading obtained on the graduated circle of the cylinder is multiplied by a predetermined constant.

To afford a wide range the windings of the main coil of the electrodynamo-meter, which is chiefly made as a wattmeter and especially suitable for calibrating meters, is wound in two halves, so that, using the suitable terminals, the two windings can be connected either in parallel or series. By means of a commutator resistances for medium pressures of 50, 100 or 200 volts can be coupled in series with the moveable coil.

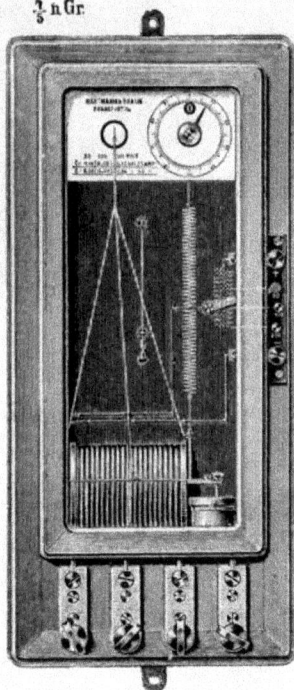

Backboard 55 × 25 cm.

The apparatus, which is suitable for use with both alternating and direct current, is mounted in a glass fronted polished walnut wood case and the measurements are made without opening the front.

No. **543.** **Electrodynamic spring balance** (Wattmeter)
I. for currents up to 50 amps. . . . . M. **250.—**
II. „   „   „ „ 200   „   . . . . „ **300.—**
Instruments for other currents and potentials specially quoted for.

No. 537 and 538. Recording Ammeter and Voltmeter for direct and alternating current.

⅕ nſtr.

These **recording voltmeters** and **ammeters** consist of a thoroughly damped Kohlrausch galvanometer in which a pen is combined with the index, and a drum driven by clockwork on which a paper strip with time divisions can be fixed. The Kohlrausch galvanometer is specially suitable for this class of work as the amount of power it exerts is sufficient to obviate any question of the friction of the pen causing inaccuracies in the readings; a screw adjustment is provided to enable the pressure of the pen on the paper to be adjusted with the greatest nicety and the core can be regulated to hang perfectly free from friction on a thin stretched wire in the solenoid by means of adjusting screws and a plummet.

The clockwork concealed in the drum is usually arranged for a 24 hours reading, but if desired, the drum can, without extra charge, be arranged to revolve once in seven days or at any other desired speed.

The cases are made in polished walnut with lock up glass fronts.

Backboard 55 × 23 cm.

No. 537. Recording Ammeters for direct current.

No. 537 and 537a*)	II	III	IV	V	VI	VII	VIII	IX	X
For amps. max. . .	5	10,20,30,50 or 75	100 or 130	200 or 250	300	400	500 or 600	700 or 800	1000
Price in M. . . .		270	260.—	260.—	310.—	320	330	340.—	380.—	475

No. 538. Recording Voltmeters for direct current.

No. 538 and 538a*)	II	III	IV	V	VI	VII	VIII	IX	X
For i. e. up to volts	50—75	90—120	50	100	150 or 200	300 or 400	500 or 600	700 or 800	1000
Price in M. . . .	275.—	275.—	275	275	300.—	325.—	350.—	375.—	420.—

Nos. **539** and **539a**). **Combined Recording Ampère- and Voltmeter,** in one case and recording on one drum; the price is then reduced from the total amount of the resp. prices of Nos. 537 and 538 together by . . . . M. 50.—

Blanks with time divisions per 100 M. **2.—.**   **Spare-Pens** each M. **5.—**

Attachment for Recording Ammeters in accumulator installations, indicating the time occupied in charging.   Extra M. **30.—**

*) The numbers marked with **a** are for **alternating current** instruments.

No. 540. Electrical Meter (Watthour Meter)
for direct and alternating current.

System patented.

These **Watthour meters** with intermittent registration consist of an electro-dynamic balance on the same principle as that described on page 117, a very simple electrically driven train of wheel work and a counting gear with electro-magnet. The train of wheels act at regular intervals on the spiral spring which opposes the motion of the wattmeter coil, winding the suspension wire thereof over a roller and afterwards unrolling same. At the moment that this action commences the counting gear is connected by an electromagnet with the roller and this connection is maintained until the spiral spring and wattmeter coil are balanced, that is until the latter has been brought back to its zero position, directly this occurs no current passes through the electromagnet and the counting gear is disconnected from the roller by a suitable spring. The amount the spring is extended and the counting gear advanced is exactly proportional to the power passing through the wattmeter.

To render the meter more compact a balance with sliding weight in lieu of the electrodynamic balance is fitted. In the alternating current pattern a hot-wire instead of the electromagnet serves to put the counting gear in action.

As there are no loose parts to be removed for travelling as well as no means of stopping the moving parts from outside the case the counter can be passed by the authorities. The various parts are mounted on a polished walnut wood base and protected by a sheet metal case which can be closed by a lead seal or otherwise fastened after calibration if desired.

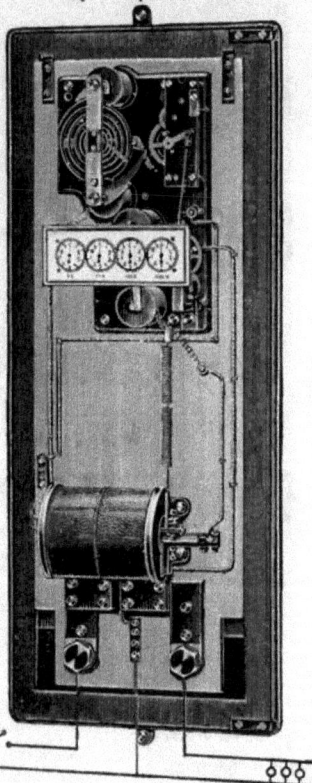

Backboard 55 × 25 cm.

No. **541.** **Watthour Meters** for direct current.

No. **541**	0	I	II	III	IV	V	VI	VII	VIII	For the three wire system the extra cost of instruments for the same current is from M. **30** to **50**.—.
Up to amps max.	15	25	50	75	100	200	300	400	500	
Price M. . . .	125	130	140	160	180	200	250	275	300	

No. **541a.** **Watthour Meters** for alternating current for the same number of ampères are supplied at the same prices as direct current instruments.

No. 544.
Electrical Time Meter for direct and alternating current.

Diameter
of base
160 mm.

Metal cover is
indicated
in the woodcut.

$\frac{1}{3}$ n. Gr.

In a great number of instances where the same amount of current is always taken from a central station the use of the more expensive meters such as ampèrehour or volt coulomb meters can be avoided by substituting a time meter which registers the number of hours during which energy is consumed. The use of such simple meters should induce a large number of users of a small but definite number of lamps to be connected to a central station supply who at present do not apply on account of the comparatively excessive cost of meter rental in comparison with their outlay for current, whilst the possibility of more current than agreed for being fraudulently obtained can be prevented by suitable fuses or preferably by the use of the automatic switch described on the next page.

These **Time Meters** are also well adapted to determine the current absorbed by motors intermittantly in circuit, to determine the amount of discount allowable for current consumed for motive purposes, to check the number of hours lamps have been lit in railway carriages, or that electrical launches have been at work.

As compared with the time meters hitherto employed with spring driven clockwork and electrical starting and stopping gear to the pendulum, these time meters have the great advantage of absolutely electrically driven clockwork which is quite unaffected by shocks or alterations of position such as occur in conveyances. The dials are graduated up to 1000 hours direct to 5 minutes, and the terminals are so arranged in the polished walnut wood base that they cannot be got at to short circuit them and so cut the meter out of circuit. The metal cover is fitted with means to insert a lead seal.

No. **544. Electrical Time Meter** for direct current at the usual working pressures*) . . . . M. 38.—

No. **544a. The same** for alternating current at the usual working pressures.

*) The working pressure is to be stated on order.

No. 527.
Automatic Double Cut out for direct and alternating current.

$^1/_5$ n. Gr.

System
patented.

Size
of backboard
22×22 cm.

In small central stations, more especially in those driven by water power and in which therefore the cost of generating the current does not vary with the demand but is always equivalent to the cost of the maximum output, current is frequently supplied for a certain contract sum instead of by meter, and the contract price is principally based on the maximum demand and not on the time it is required per day. The use of the main switch here described is advisable to prevent users exceeding the stipulated maximum current by extending their installation without notice.

This self acting **Double Cut out** comprises a spring galvanometer in which the core acts on one arm of an excentrically pivotted lever, a flat spring strip bearing against the other arm; this strip is fitted with a catch which keeps the switch block between its spring contacts, if the iron core is drawn so deeply into the solenoid that the flat spring looses its support thereby disengaging the catch from the switch block the latter flies back under the action of a spiral spring. Another lever is provided by which the switch block can be replaced in position after the cause for the increased demand for current has been removed; this lever is so arranged that it is impossible to put the automatic action out of gear by its use.

The apparatus is contained in a polished walnut wood case and is also an excellent substitute for the main fuses, it can be adjusted, in position, within fairly wide limits, locked and lead sealed, and works with certainty within $\pm$ 1%.

No. 527. Automatic Double Cut out for direct current.

No. **527** and **527a**	I	II	III	IV	V	VI	VII	VIII
For amps. max. . .	2	5	10	15	20	25	30	40
Price in M. . . .	45.—	44.—	42.—	42.—	42.—	44.—	46.—	50.—

No. **527a**. **The above switch** for alternating current at the same price.

1/4 n. Gr.

Detector Galvanometers.
No. 364.
Vertical Detector.

No. 364. **Vertical detector** in polished mahogany case with semicircular magnet resting on knife edges the magnet poles projecting into coils, it can be used in any position as regards the meridian and is fitted with travelling stop and controlling magnet. Resistance about 10 ohms.

Sensitiveness 1° deflection = approx. 0.00004 amp.

Price M. **32.—**

No. 365. Vertical astatic Galvanometer.

1/5 n. Gr.

No. 365. **Vertical astatic galvanometer**. (Differential) Braun's type. A horse shoe magnet made from steel tube is mounted at one end of a very sensitive balancebeam with its poles projecting into two solenoids the winding on which can be coupled externally as desired; it is quick in action owing to a very effectual air damper fixed at the other end of the beam. Safe travelling stop. For use in any position as regards the meridian. Total Resistance about 100 ohms. Sensitiveness 1° deflection = approx. 0.000005 ampère. Price M. **65.—**

No. 515. Vertical Detector.
(German Post Office type.).

1/3 n. Gr.

No. 515. **Vertical detector** with ∧ shaped magnet mounted in centres on palisander wood base with solid brass case and scale etched on glass, in two forms:

I. Of about 20 ohms resistance. Price M. **18.—**
II. Of about 700 ohms resistance. Price M. **25.—**
Sensitiveness with the higher resistance: 1° deflection (1/10 division) = 0.00006 ampère.

No. 516. Horizontal detector
(Swiss type)

1/3 n Gr.

much used in telegraph work; ∠7 shaped magnet supported on needle point, with travelling stop, for convenient adjustment to the meridian can be turned in the mahogany base. Two turns of thick and several turns of thin wire of about 1 ohm furnish two degrees of sensitiveness. With the latter in circuit 1° deflection represents approx. 0.0001 ampère.

Price M. **45.—**

No 319 a

No 514

No 319 c

No 319 b

No 380 a

No 381

No 380

Volt

No 383 a

No 383

No 381 N

No 382

No 381 c

No 381 r

No 379

No 513

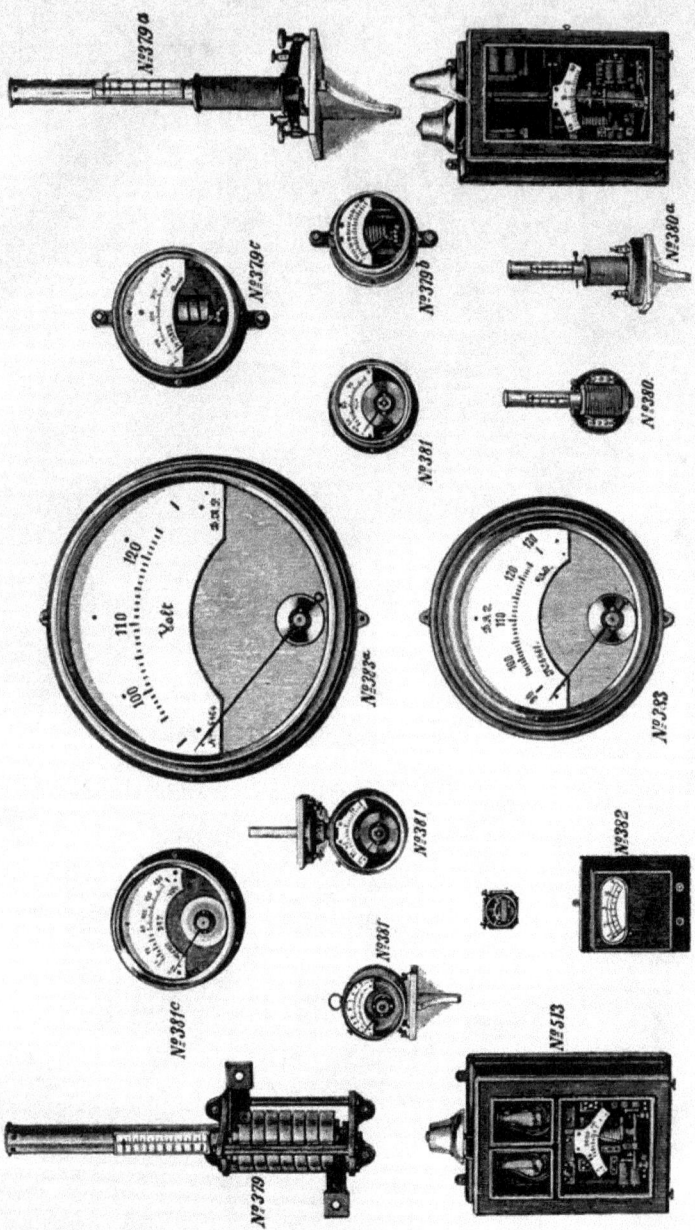

Comparative Sizes of Volt- and Ammeters.

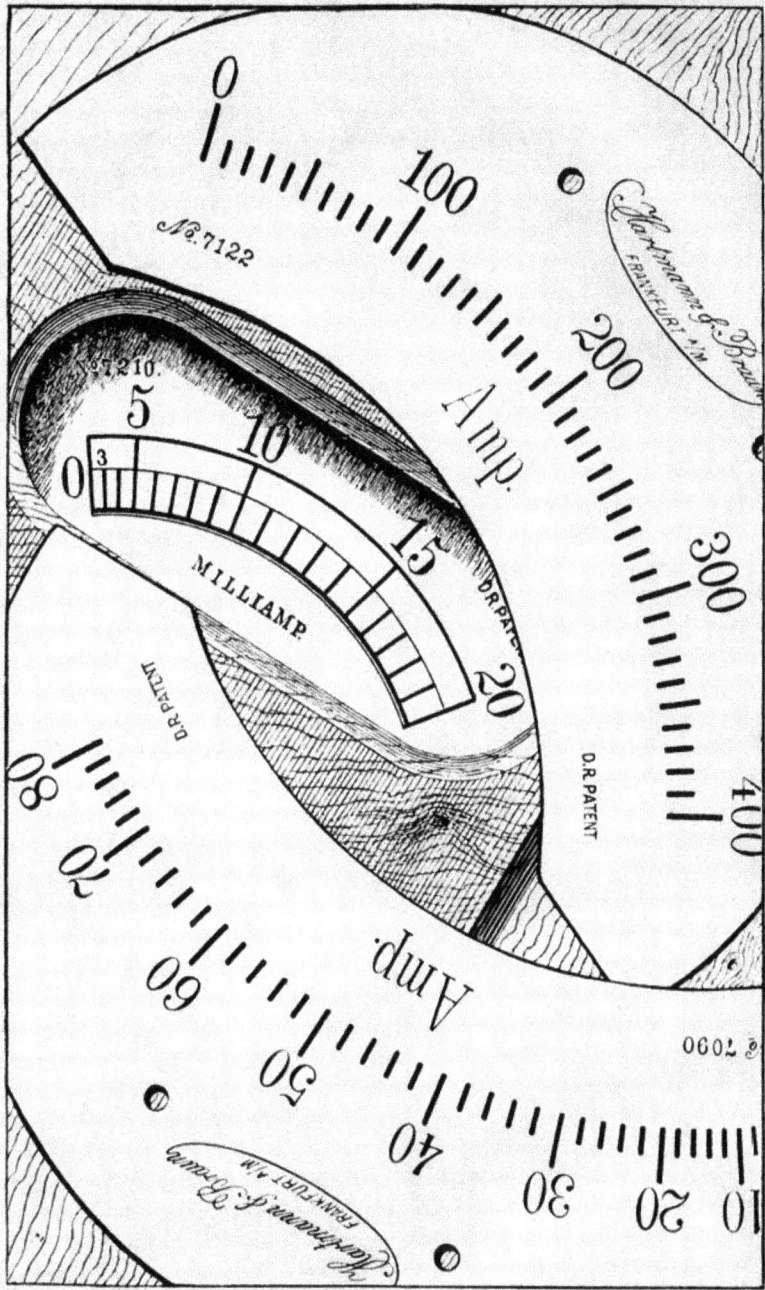

Specimens of Scales of electromagnetic Ammeters (pages 102 and 105).

126

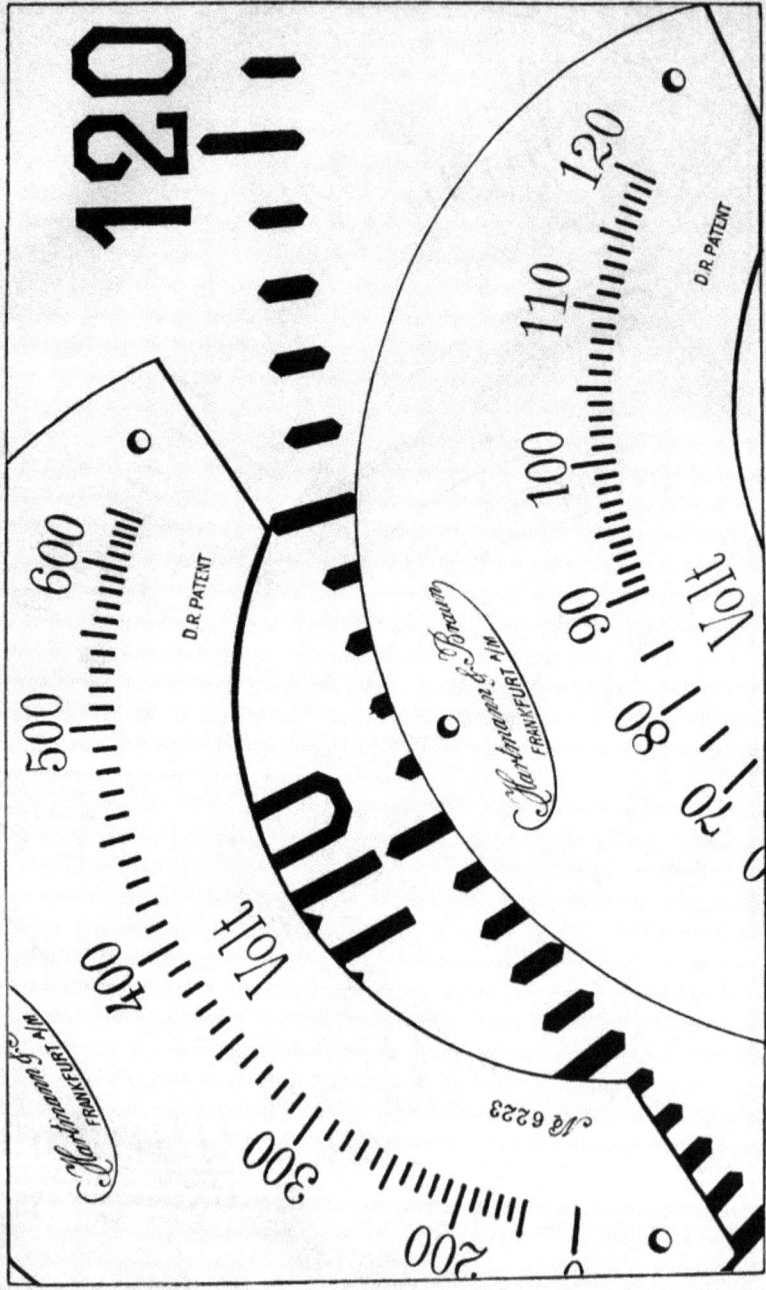

Specimens of Scales of electromagnetic Voltmeters (page 103) and of the Universal Voltmeter (page 107).

XI. Switches.

No. 520. Commutator switch to Switch in series or parallel.

No. **520.** This **Commutator** is for use in accumulator installations to couple two halves of a battery in parallel when charging, and when charged, in series. It is also suitable for changing the source of current in measurements, and inserting shunts in the circuits of instruments. Mounted on lacquered cast iron base. For currents up to 100 ampères.

Price M. **60.—**

Automatic Switches of various designs.

Apparatus of this class are made for the most diverse purposes and sometimes to carry very large currents, such as an automatic switch to insert a shunt to carry 1000 ampères with a recording ammeter for 400 ampères. The release of a weight acting by gravity to insert the shunt between the contact springs with a total contact surface of 10000 square millimetres is effected within an accuracy of 1—2 per cent on attaining the maximum current in the measuring instrument.

This and other similar classes of switching apparatus specially quoted for.

Another automatic switch has already been described on page 123. This **Double Cut out** when used instead of lead fuses has the advantage that when the circuit has been broken through overloading it can be at once remade and further that the break occurs if the desired maximum current is exceeded by 1—2 per cent. The extra cost of this switch is hardly worth consideration when compared with the advantages it offers, and this extra cost is quickly saved by the economy in material and labour arising from not having to replace blown fuses.

No. 521 and No. 521a. Automatic Switches (cut-outs) for maximum and minimum current.

⅛ n. Gr.

No. **521**.   No. **521a**.

These automatic cut-outs, intended chiefly for use in accumulator installations, are mounted on lacquered cast iron bases. The switch lever is fitted with a copper blade which works sweetly in contact between several copper spring strips of which the superficial area is large enough to ensure very slight friction in the contacts at the rated carrying capacity, a catch keeps it in position between the contacts whilst a spiral spring withdraws it.

No. **521**. **Automatic cut-out for maximum current.** The release of the catch holding the lever in position, when the maximum admissible current is exceeded, is effected by the movement of the armature towards the electromagnet; this can be adjusted to within 2 per cent for any desired maximum within the ranges given below.

No. **521a**. **Automatic cut-out for minimum current.** The permanently magnetised steel armature falls directly no current is passing through the electromagnet or one is passing in the reverse direction, and in falling releases the catch holding the switch lever in position.

Both types are supplied at the same price to carry the following currents.

Nos. **521** and **521a**	III	IV	V
To amp.	20, 30, 50 or 75	100 or 150	200 or 250
Price in M.	**75.—**	**75.—**	**95.—**

Simple Cell Switches.

No. **523.** **Cell switch** with screw traverse mounted on lacquered cast iron base; the sliding contact has spring contacts which are fitted with intermediate resistance, it has sufficient area to carry the maximum current and moves in a straight line. It is actuated by a quick pitched screw which can be used either horizontally or vertically. Besides taking up less room this arrangement enables the leads to the various cells to be arranged in the most convenient manner.

I For currents up to 50 ampères

for	6	9	12	15	cells.
M.	85.—	90.—	100.—	120.—.	

II For currents up to 100 ampères

for	6	8	10	cells
M.	90.—	100.—	135.—	

For any intermediate number of cells the next higher price is charged.

Double cell switches. Prices specially quoted.

No. **524. Instantaneous cell switch** in which pressure on lever *B* causes the rubbing contact mounted on a spring case with toothed wheel to advance one tooth i. e. on to the next contact thereby putting another cell in circuit. Switching off is effected by turning the lever *A* backwards. In this arrangement it is impossible for the rubbing contact to remain in position between two contact plates.

⅕n. Gr.   A   B

For currents up to 100 ampères for 7 cells . . . . . . M. **55.**—
For each extra cell, to 11 cells . . . . „ **3.**—

Automatic Pressure Regulators.

⅛ n. Gr.

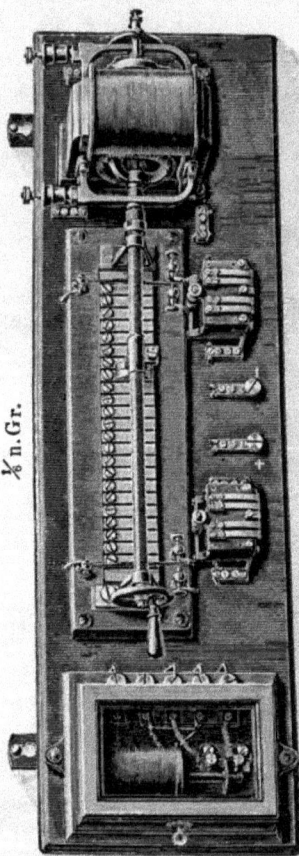

No. 525.
Battery Regulator Switch.

No. 526.
Shunt Regulator Switch.

No. 526a.
Main current Regulator Switch.

The **automatic pressure regulators** are intended to maintain a constant pressure in a system of mains. For this purpose the apparatus is fitted with a sensitive contact voltmeter No. 510, which, when any variation of pressure occurs in the network, drives, through two relays No. 511, a small electromotor in one direction or the other. The motor is connected to a similar switch arrangement to that described in No. 523, by which means cells or resistances are inserted or cut out from the shunt circuit of the dynamo or the main circuit until the pressure has again reached the normal amount.

The various instruments are mounted together on a polished walnut base and protected by a glass fronted case of the same wood. In many cases it is more convenient to separate the contact voltmeter and relay from the motor and switch, in which case the instruments Nos. 510a and 511b in circular cases can be used.

No. 525. Automatic Battery switch

| for 6 | 9 | 12 | 15 cells | to 50 amps. | M. 590.— | 600.— | 630.— | 650.— |

to 100 amps. | for 6 | 8 | 10 cells |
| M. 600.— | 630.— | 670.— |

For an intermediate number of cells the next higher price is charged.

No. 526. Automatic shunt regulators for currents up to 15 amps. with 20—24 contacts . . . . M. 630.—

No. 526a. Automatic main current regulators for currents up to 100 amps. according } M. 670.— to 750.—
to the number of contacts . . . . (maximum number 30)

Resistances for shunt and main current regulators are specially charged for.

XII. Apparatus and Material for Wiring-contractors and -inspectors.

No. 621.
Portable Ammeters, Voltmeters &c., for wiremen and inspectors.

Size of portable case:

25×25×12 centimetres.

To enable the cheaper electromagnetic instruments shown on pages 102 and 103 or any other instrument mounted in circular case with base of 225 mm diameter, such as the Leakage Indicator page 117 or the Wattmeter page 118 to be used as portable instruments for installation purposes, we supply them mounted in solid oak cases with handle. These cases are fitted with lock up fronts opening downwards so that the scale can be easily read, and to render the comparatively narrow cases firm in use without affecting their portability the bases are fitted with adjustable metal feet, one of which has a regulating screw attached with which the pointer can be brought to zero on the scale.

Portable case for A m m e t e r, not suitable for currents exceeding 150 ampères. The terminals projecting beyond the circular case are removed and others projecting through the back of the wooden case are substituted; the coupling screws are provided for inside same.

Price M. **30.—**

No. **621a.** **Portable case** for V o l t m e t e r. Price M. **25.—**

Other portable instruments,

on the Deprez principle see page 94,
" " caloric " " " 98,
" " electromagnetic principle " " 104.

No. 384c. Insulation Tester for Wiremen.

⅛ n. Gr.

Size
of case:

G   W

24 × 20 × 18
centimetres.

In testing the insulation resistance of small installations such as house wiring, the use of apparatus fitted with a magneto inductor is far preferable to those depending on a few dry cells to furnish the requisite current, as the E. M. F. of the inductor approximates the pressure at which the installation will be worked, whilst cells occupying the same space would not furnish more than 20 volts and would gradually decrease in power; the inductor on the contracy remains constant.

The following apparatus contain magnetos for about 100 volts, partly with bell, partly with galvanoscop, in solid oak cases with handle; the crank is to take off and to place inside the case.

No. **384a. Simple Insulation tester** with b e l l s. The apparatus contains an alternating current magneto which rings the polarised bells through a resistance of 25—30000 ohms.   Price M. **50.**—

No. **384b. Insulation tester** with d e t e c t o r to approximately determine the amount of resistance in ohms; comprising a direct current magneto, a detector No. 515 page 124, and a comparative resistance of 20000 ohms which can be inserted by means of a switch. The detector is calibrated to indicate correctly when the handle is turned at the rate of not quite 3 revolutions per second (the military quick pace*) — and measures resistances from 10 to over 100000 ohms within 5000 ohms. Price M. **75.**—

No. **384c. Combined Insulation tester** with b e l l s and d e t e c t o r. The two instruments above mentioned are here combined in one with a magneto for both alternating and direct current; two pairs of terminals are therefore fitted of which one pair *W* is for testing insulation by sounding the bells and the other pair *G* for use with the detector. Price M. **100.**—

*) Experiments have proved that most people are able to maintain this speed within a few per cent. The comparative resistance affords a ready means of determining the correct speed, as when switched into circuit whilst turning the magneto the pointer of detector must show **20000** ohms.

No. 624. May's Watch shaped Speed Counter with measuring wheel.

The great advantage of this speed counter, other than its extremely convenient shape, is that the figure dial automatically changes to correspond with the direction in which the shaft is rotating. As only the correct set of figures corresponding to the direction of rotation can be seen it is impossible for errors in reading to arise. Two hands similar to those used in a watch show the number of revolutions of the shaft. The long hand marking units and tens and the small hand the hundreds. The instrument is easily reset by opening the back of the case and turning a small milled nut.

The triangular point can be quickly replaced by a small disc with rubber let into the circumference and exactly 20 cm round, to measure the amount of cable employed, belt speeds &c., the readings being divided by 5 to give the result in metres.

No. 624. Speed Counter with **metre measuring wheel,** in case.
Price M. **25.—**

The first wood cut shows the counter after use against a shaft revolving in the same direction as the hands of a watch, in other words, to the right, the second after revolving to the left; the reading in both instances being **738**.

A milled edged nut which is accessible on opening the back of the case enables the counter to be easily and quickly reset to zero.

No. 625. The same **Counter** with **Chronograph** (Second-watch).
Price on application.

No. 627.

Mechanical apparatus to determine the sectional area of mains or the fall of potential therein.

Helberger system.

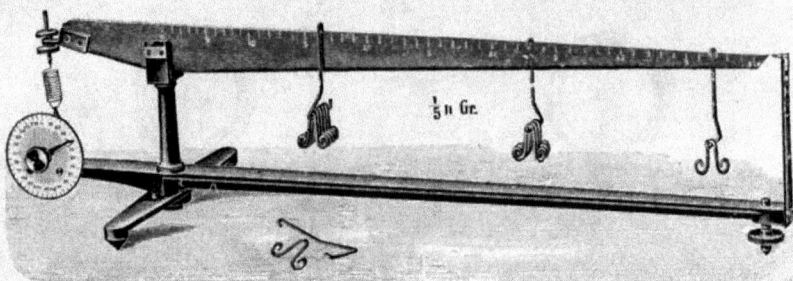

This apparatus furnishes wiremen with a ready means of ascertaining mechanically the correct sizes of the leads in electrical installations, especially the requisite sectional area of the mains in house wiring — the sizes for branches need not as a rule be determined — and is also very useful to work out the feeder network for central stations in fact for all installations supplied from one source.

It consists of a two armed lever in which the long arm is graduated and the short arm is fitted with a spring balance and pointer.

To use it, weights — proportional to the number of lamps or amount of current — are hung on the graduated arm at a distance from the bearing corresponding to the actual distance of the lamps from the source of supply, and the spring is then screwed up until the lever floats opposite the beam pointer. The index on the divided circle then shows the sectional area or diameter of the wire practically suitable, allowing for the fall of potential admissible, the latter can also be determined for any desired sectional area without further manipulation.

No. 627. **Apparatus to determine the sectional area of mains,** in a flat wooden case, with 50 weights. Price M. 85.—

No. 628.

Apparatus to determine mechanically the sectional area, drop in potential, and loading in a network of mains. Helberger system.

Patented.

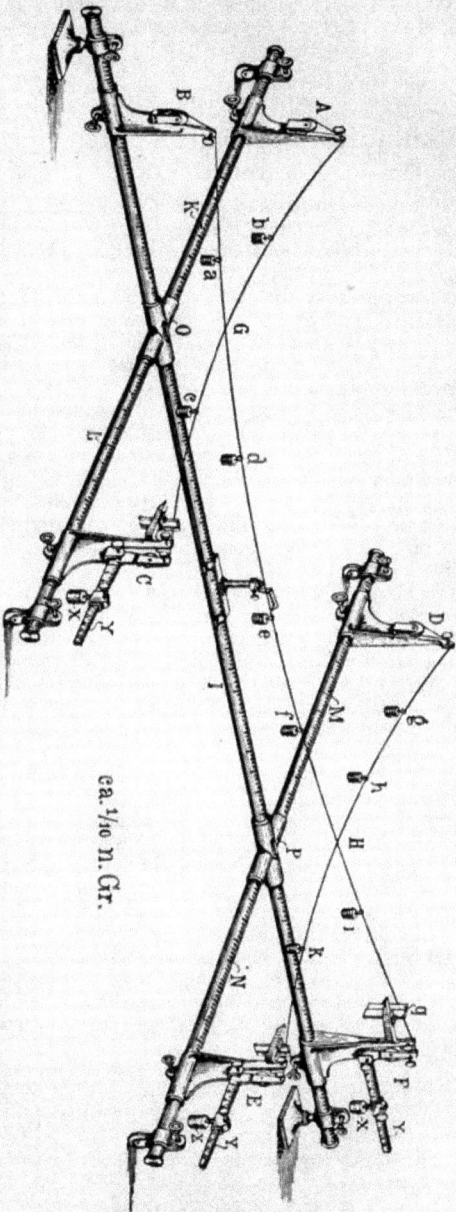

ca. 1/10 n. Gr.

AC, BF, DE represent distributing mains A, B, C, D, E, F feeding points to same,
a, b, c, d, e, &c. are points where current is taken off, the amounts being represented by weights.

No. 628.
Cable Calculating Apparatus. Helberger system.
Patented.

This cable calculating apparatus enables the sectional area, drop in potential and distribution of current in a network of mains fed at various points to be determined in the most simple manner without any complicated calculations.

Experiments proved that the behaviour of a thread suspended at two points and loaded with weights is the same as that of a wire to which current is supplied at its two ends and from which current is taken at several intermediate points, and this fact has been adopted in designing this apparatus for calculating cables.

It is built up chiefly of several cylindrical graduated rails which can be connected crosswise together. On each of these rails two moveable supports are mounted which carry the other parts necessary to adjust the threads and read off the results, a number of weights and an arrangement to determine the dip of the thread are also supplied. These parts correspond to the factors for a distribution network of length and junctions of the mains, loading and connections thereto and also for the fall in potential admissible.

The moveable supports are arranged on the rails so that their relative distances apart correspond with the lengths of the mains between the feeding points, they are then connected together by threads on which weights proportional to the number of lamps or current supplied are hung at distances proportional to the relative distances of these delivery points. The sliding weight on the lever carried by the support is now adjusted until the dip of the thread corresponding to the drop in potential nowhere exceeds a given amount.

The position of the sliding weight then affords a measure for the sectional area of main necessary whilst the declination angle of the thread from the horizontal indicates the distribution of the total energy.

The whole apparatus is designed to render all manipulation as simple as possible, each support for instance being fitted with a thread holder in which the thread is automatically wound up when not in use. The declination angle of thread is read off direct on a small scale mounted on the lever support. The dip can be adjusted for various losses of potential and, like the weights, is calibrated from definite mathematically tested units.

The apparatus, owing to the amount of tedious calculations obviated thereby, is especially suited for use in a engineer's office, and also for technical schools to experimentally demonstrate the application of Kirchhoff's law to calculations for central stations.

No. 628. **Cable Calculating Apparatus** with one longitudinal and two cross rails and 6 supports; the rails made of brass tube.

Price M. **600.**—

No. **628a. The same apparatus** with the rails in wood „ **510.**—

Extra **cross rails,** each with two standards
for No. **628** M. **170.**—; for No. **628a** M. **135.**—.

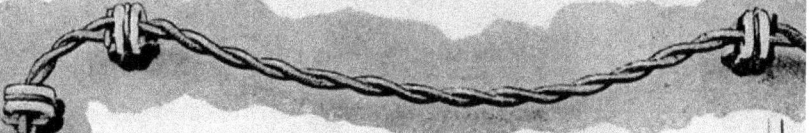

Ring insulators,

protected by patents and registered designs.

A new system for wiring inhabited rooms was worked out last year in our offices, which has proved very successful and has to a very great extent been adopted by all large wiring contractors.

It has been found everywhere where central stations have been erected that those people whose houses are best decorated are those whose custom it is most difficult to obtain and therefore the central station looses or rather does not obtain the customers who would consume the largest amount of current. The cause for this is to be sought in the fact that, if the wiring is carried out in accordance with the rightly somewhat severe rules necessary to ensure durable work, it offends the eye, or, if concealed behind the plastering, the making good of walls and ceilings causes in any case great mess and disturbance.

The ring insulator system meets the requirements of the electrician, the architect and the client as it offers every advantage for permanently satisfactory work, can be matched to any style or tone of decoration and can be installed without causing any dirt and at a considerably less cost than the other systems usually adopted.

The materials necessary for installations on this system are made in two sizes and a large stock of same is kept and also of the requisite special tools.

All packages bear the trade mark illustrated at side.

Full instructions in pamphlet form and **price lists** will be sent on application.

The chief parts of this ring insulator system are small rings of porcelain or glass with a slot round the edge which are sprung into brass hooks (plain, nickelled, or stained blue or brown).

nat. Gr.

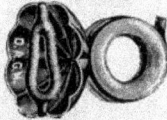

nat. Gr.

The hooks are ornamented with embossed rosettes which are either suspended on round headed wood screws, a method chiefly adopted in hired dwellings, and hospitals or for temporary purposes, or else are nailed up using always small wood plugs to fasten into the walls.

Special tools are provided to render this simple method of installation as easy as possible. Two blows with a heavy hammer on a conically pointed round steel chisel make a suitable hole

nat. Gr.

for the wood plug even in the hardest brickwork. If the chisel is carefully removed not a particle of dirt or dust is caused. Another hammer blow drives the dry wood plug into the hole thus made using no plaster whatever except in the case of extremely rotten plaster work. Then, using the seating tool, the nail hook is gently hammered in, the rosette hiding the wooden plug.

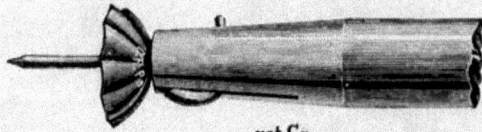

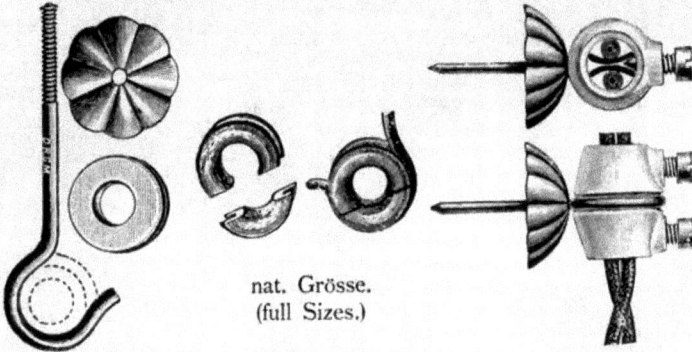

nat. Grösse.
(full Sizes.)

Screw hooks in five lengths from 15 to 150 mm are provided to carry leads from the wall across the ceiling to a chandelier.

The class of wires most suitable are flexible cords or stranded wires covered, according to the class of room with either silk or worsted to match the colour of the paper or other decorations. Specially shaped wedges of insulating material serve to clamp the wiring in the insulator, whilst particular forms of gripe insulators are provided to take the strain of the wires off the switch or cut-out terminals.

Split rings can be used with advantage when subsequently inserting new supports or using very heavy long leads.

Improvements in all the other work incidental to wiring have also been made, for instance the flexibles are neatly ended off and joints made without the dangerous use of solder by using thin metal rings (eyelets), the bare ends of the wires are bent into these rings and their edges are closed over and flattened by suitable pliers; wires finished off with these ends are connected together sometimes by hollow rivets and sometimes by a special type of screw rivet.

Special wall drills and attachments are also made enabling the thickest walls to be pierced without noise or dirt and leaving a clean cylindrical hole.

Using these ring insulators and special incandescent lamp holders with external contact terminals, gas chandeliers can be easily and neatly converted for electric light.

XIII. Telephones.

No. 660a.

Station with microphone and battery call.

In large telephone installations or in smaller ones where the instruments are in constant use it is preferable to erect instruments which although dearer than the usual class of telephones are however of far better workmanship and made to stand hard work without damage, of this class more than 80000 sets have been supplied to and are now in use by the German Post Office who have been constant customers of our firm for many years.

The above apparatus with carbon pencil microphones in which the pencil vibrations are controlled by springs and with watch shape telephones of our own design are supplied for use either with battery or magneto calls. The cases are made in polished walnut wood and all the metal work is made and fitted with the greatest care and accuracy.

No. **660. Station** with one telephone for battery call.

No. **660a.** As above, but with two telephones.

No. **662. Station** with one telephone and magneto call.

No. **662a.** As above, but with two telephones.

Prices of these instruments which are not supplied singly, but only in quantities of not less than 50, will be quoted upon application.

The batteries for call and microphone have to be mounted in separate cases and are not supplied with the instruments.

XIV. Small Dynamos and Motors.

No. 638.
Portable Dynamo
for
blasting purposes.

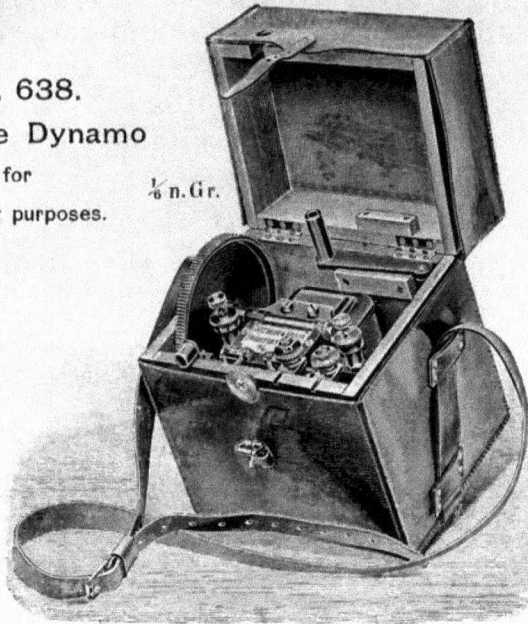

⅛ n. Gr.

This portable dynamo is specially intended for blasting purposes using incandescent fuses and as compared with the use of static electricity has the advantage that it is absolutely independant of the weather and that failures therefore can never arise. It is however suitable for all purposes where a current at a high potential for a short space of time is required. Special attention has been paid to designing a machine as light in weight as possible and of the maximum attainable output. The weight is approximately 10.8 Kilogrammes and at 3 revolutions of the handle per second the machine gives over 1 ampère with an output of about 50 watts.

No. **638. Portable Dynamos** with switch mounted in solid oak case, with strong leather case and carrying strap. Price M. **340.—**

Revolutions per minute	Potential in volts	Current in ampères in an external resistance of ohms				Smallest external resistance admissible in ohms	
		100	75	50	25		
2000	19.5	0.20	0.26	0.39	0.78	16	*) In practice the maximum current admissible is 1.2 ampères.
2500	37.5	0.38	0.50	0.75	1.50*)	31	
2700	42.0	0.42	0.56	0.84	1.68*)	35	

Incandescent fuses of platinum wire 0.05 mm dia. (a lenght of about 5 mm = 5 ohms) require about 0.6 ampère to ensure firing.

No. 630. Electromotors for direct current.

$\frac{1}{5}$
n. Gr.

In designing the larger sizes of electromotors special attention is always paid to attaining the highest possible ratio of efficiency whilst this point has been mostly disregarded in the smaller types. In these latter the efficiency depends chiefly on perfect workmanship in the mechanical details and their manufacture is therefore more suited to a mechanician's than an engineer's workshop. A comparison between the efficiencies mentioned below and those of other small motors will justify this statement.

The details of construction in the motors here illustrated have several special features, the fields are of the horse shoe type with ring armature, they are absolutely sparkless even if overloaded and the brushes do not require adjustment, the commutator is of hard metal and in conjunction with the round carbon brushes affords every guarantee for continuous satisfactory running in either direction. To enable the motors to run in any position grease lubrication is adopted for the bearings. Like all other small motors these are wound as series machines, the speed therefore varying with the load.

No. 630. Electromotors.

No. 630.	Output in H. P.	Watts absorbed	Efficiency in %	Revolutions per Minute	Diameters of the 3 speed grooved pulley in mm	Space occupied in cm			Total weight kg	Price in M
						Length	Breadth	Height		
III	$\frac{1}{20}$	66	55	1700	40, 50, 60	20	15	18	6,5	165.—
IV	$\frac{1}{15}$	81	60	1700	40, 50, 60	20	20	21	12,5	170.—
VI	$\frac{1}{8}$	131	70	1500	55, 80, 100	23	28	28	30,0	260.—

Resistances to regulate the speed specially quoted for.

No. 632. Electrically driven Ventilating Fans.

ca.¹/₆ n. Gr.

The inconveniences attached to working ventilating fans in houses and especially in workshops, due especially to the high speed at which they require to run, to a great extent prevent their use. This latter point however is one specially adapted to be dealt with by electrical means where available, and where fresh air or the removal of hot air or noxious fumes is required. Electromotors owing to the small space they occupy, easy starting and slight attention required are extremely suitable for driving these fans Both motor and fan are mounted together on a cast iron bracket and the iron tube which is built into the wall is concealed by a moulded frame. All parts are highly lacquered.

No. 632. Electrically driven Ventilating Fans.

No. 632.	III	IV	VI
Watts absorbed . . .	66	81	131
Diameter of fan . . .	250	300	350
Price in M. . . .	**210.—**	**215.—**	**320.—**

Resistances to regulate the speed specially quoted for.

XV. Electrical Thermometric Apparatus.

Prof. Braun's Standard Pyrometer.

The Braun Pyrometer for measuring high temperatures up to 1200° Celsius, is based on the alteration in resistance of a platinum wire in accordance with the temperature of the chamber in which it is inserted. This platinum wire which is wound non-inductively on a fire clay former and mounted in a fire proof tube forms one arm of a Wheatstone bridge in connection with a sensitive galvanometer by means of which the existing resistance can be easily ascertained or in other words temperature read off direct in degrees.

The actual measuring apparatus is fitted into a strong lock up oak case and so arranged that only those parts requiring adjustment are accessible, the cells are mounted in a separate compartment closed by a cover at the back of the case, the galvanometer is also fitted into a separate case for travelling purposes, can be used in any position and is unaffected by external magnetic fields.

The working of the apparatus is extremely simple and can be carried out by any unskilled person.

The instrument is sufficiently sensitive to render an increase or decrease of a few degrees of heat, even at the highest temperatures, distinctly perceptible. The degree of absolute accuracy is extremely high as the instruments are calibrated in connection with a great number of readings taken with the porcelain air thermometer.

The pyrometers are supplied graduated to either 1000 or 1200° Celsius the measuring apparatus in both patterns is the same but the construction of the pyrometer is essentially different.

No. **480a. Pyrometer bridge** with a galvanometer easily set up for use in any position, calibrated as desired to either 1000 or 1200°.

Price M. **425.—**

No. **482. Pyrometer** in iron tube fitted with pure nickel cover, and protected by an **iron sheath** lined internally with thin platinum sheet, suitable for temperatures up to 1000°.

Price, exclusive of the platinum lining M. **240.—**

The **platinum lining**, weighing about 75 grammes, is charged extra according to the market price of platinum.

No. **482a. Pyrometer** in iron tube fitted with pure nickel cover, covered at the base by a **thick platinum sheath** specially protected against external damage, suitable for temperatures up to 1200°.

Price exclusive of the platinum lining M. **240.—**

The **platinum lining**, weighing about 200 grammes, is charged extra according to the market price of platinum.

Soft flexible wire, about 3 sq. mm. area, for a distance of 10 metres, with solid ends.

Price M. **10.—**

Instructions for use: After inserting the Pyrometer into the chamber of which the heat is to be tested until the lower flange of the instrument projects therein, its terminals are connected by stout wires or cable to the two large terminals on the left side of the apparatus, the galvanometer being connected by thinner wires to two smaller terminals on the right hand side of the apparatus.

To take a reading the lever projecting at the side is drawn forward to connect the battery and galvanometer, and the disc adjusted untill the galvanometer is not deflected, The marks engraved on the scale then give without any reckoning the temperature of the pyrometer in degrees Celsius, a special feature of this pyrometer as compared with other similar instruments. The entire adjustments can be made without opening the case.

The distance between the pyrometer and the measuring apparatus can be any desired amount, the use of the instrument as a **Telepyrometer** offering no great difficulty. Data for the necessary sectional area of the cables or leads for measurements at a distance will be furnished on receipt of particulars showing the proposed distances.

Apparatus for measuring high temperatures.

Prof. Braun's Standard Electrical Pyrometer.

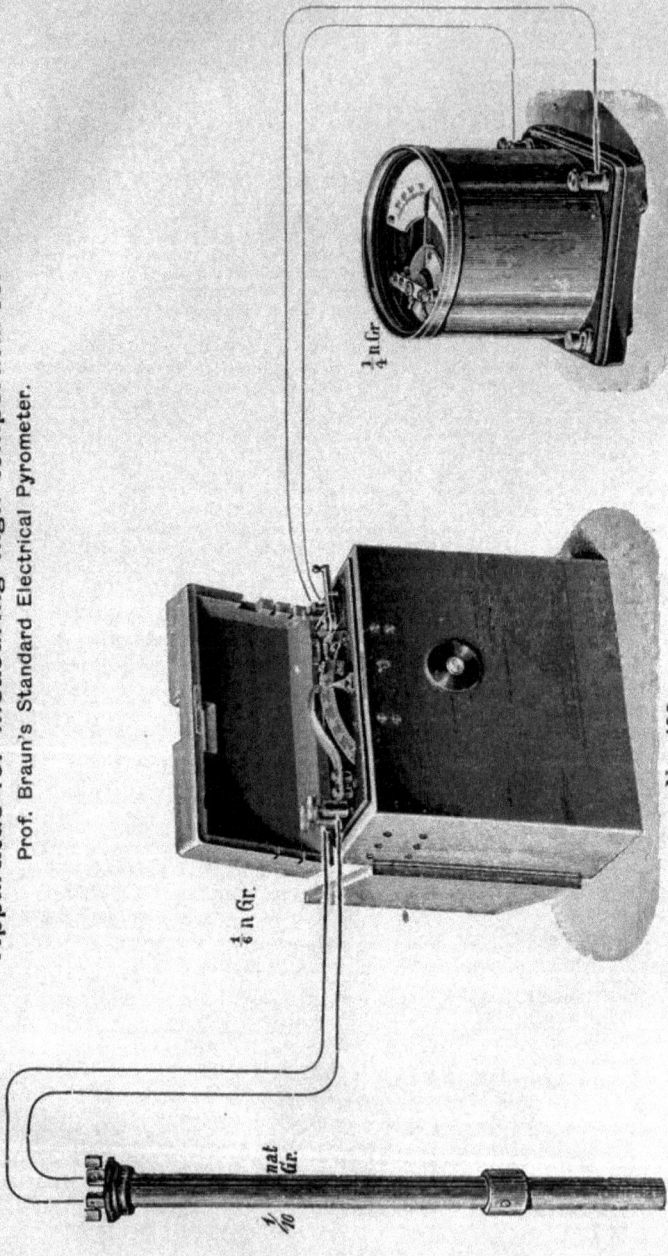

No. 536a.

$\frac{1}{4}$ n Gr.

No. 480a.

$\frac{1}{6}$ n Gr.

No. 482
or No. 482a.

$\frac{4}{10}$ $\frac{nat.}{Gr.}$

Telethermometer.

The **Telethermometer** is an apparatus easily used by unskilled hands to determine the temperature in distant rooms such as class rooms or offices which are all heated from one source, in greenhouses or inaccessible rooms such as malting floors, drying ovens, especially in pottery works, in enamelling ovens &c. The measurements can be made either in the stoke hole, or, if it is to be used as a check on the attendant, without leaving the office. Like the Braun Pyrometer it depends on the electrical resistance of a metal varying in accordance with the temperature.

Thermometer No. 485 consists of a bifilar winding on a very slender insulated polygonal mould which is protected by a perforated iron case, this form responds quickly to the slightest variation of temperature and can be used wherever the atmosphere is not too damp or contains too much acid fumes. In the latter case and for use in liquids such as in a mash tub, the thermometers No. 486 which are hermetically sealed in metal cases are preferable.

The Indicator or measuring apparatus for the telethermometer is mounted, together with a sensitive galvanometer which is uninfluenced by the magnetic meridian or by external iron, in an oak case to fix against the wall, and the temperature scale engraved on the circular dial is protected by a glass front. The measuring battery is contained in a separate case which can be placed wherever most convenient. The thermometer and measuring apparatus are connected together by well insulated wires fixed side by side.

Only one measuring apparatus is necessary to read off the temperature existing in several places as the measuring wire of each thermometer can be made to correspond in several instruments, in this case a separate lead is taken from each thermometer to the measuring apparatus whilst all can be connected to a common return wire.

The apparatus is supplied to measure temperatures from 20° below to 300° above zero in the following patterns.

No. **484. Telethermometer** with 25 contacts, to measure 25 single degrees or

50 degrees reading from 2 to 2 degrees (e. g. from 30 to 80) or
125   „   „   „   5 „ 5   „
including battery in separate case.   Price M. **285.—**

No. **484a. Telethermometer** with 50 contacts, to measure

50 single degrees (e. g. from —20 to +30°) or
100 degrees reading from 2 to 2 degrees
including battery in separate case . . . . Price M. **340.—**

No. **485. Thermometer** in perforated case for temperatures up to 150 degrees . . . . each M. **25.—**

No. **486. Thermometer** in closed case for temperatures up to 150 degrees . . . . each M. **30.—**

No. **486a. Thermometers** for temperatures up to 300 degrees
                                             each M. **36.—**

Plug or lever Commutator switch for 2, 3, 4, 5 &c. to 12 thermometers.

Price M. **10.—**, **15.—**, **20.—**, **25.—** &c., to M. **60.—**

Instructions for use. The removeable key *S* is inserted into the hole in the centre of the disc and the pointer *Z* slowly turned over the circle of metal contacts until the galvanometer needle *G* is brought back to its zero position, the temperature is then read off opposite the pointer *Z* direct in degrees Celsius. When the measurement is completed the withdrawal of key *S* breaks the circuit.

Teletethermometer-Installation.

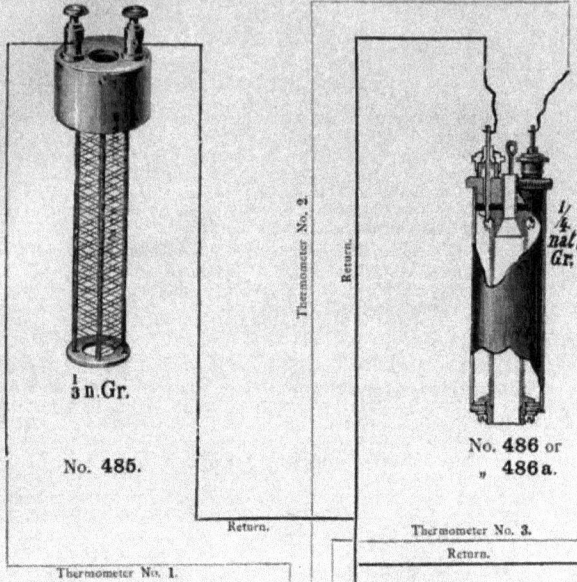

$\frac{1}{3}$ n. Gr.

No. 485.

$\frac{1}{4}$ nat. Gr.

No. 486 or
„ 486 a.

Thermometer No. 2.

Return.

Return.

Thermometer No. 3.

Return.

Thermometer No. 1.

The engraving shown an apparatus with 50 contacts, suitable for temperatures from 20 to 120 degrees, reading each 2 degrees with three thermometers, it has therefore a plug switch for the three lines.

The leads are of well insulated wire 1,5 to 2 mm sectional area of copper and any large variations in length of the various lines is equalised by inserting additional resistances between the terminals and the plug switch.

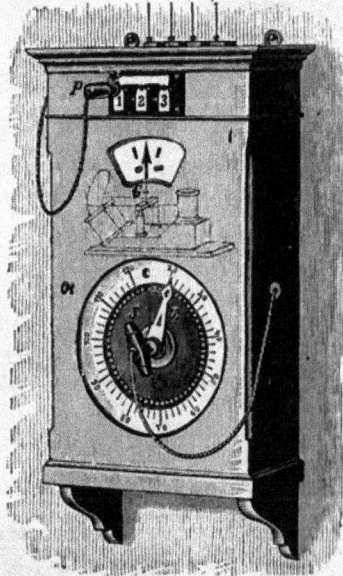

No. 484a.

Size of case:
63 × 35 × 15 cm.

XVI.
Collection of simple electrical instruments for Schools.

We have been led to design this collection of cheap electrical measuring instruments by the many enquiries we have received for the simplest class of apparatus of mechanically perfect construction, to demonstrate experimentally Ohm's law which, we are pleased to say, has latterly received attention in the science classes of schools; thereby inculcating at an early period a knowledge of the electrical units, which from the many commercial uses to which electricity is now applied, is of great advantage.

Many of the instruments employed in teaching Physics and especially in teaching Electricity are designed on a principle which the pupil cannot understand and which

No. 580.
School-

Galvano-
meter.

$1/4$ n. Gr.

in some instances is not clear to the master, with the idea we presume, of rendering them extremely sensitive or universally applicable; very often too they are made in such small sizes that their method of working cannot be seen by scholars sitting some distance off or even to render it possible for them to note the position of the index pointer, frequently too the instruments are elaborately got up and such unsuitable colours employed, that it is extremely difficult for an onlooker to recognise the actual working parts.

In the following collection we have attempted to meet the want of instruments free from the above defects. Every instrument is constructed in the simplest possible manner and arranged that every working part can be seen from a distance and especially that the scales are very clear and easily read. At the same time the instruments can be used with a comparatively great degree of accuracy and are therefore available for the teachers own private use, for which purpose some of them are fitted with additional finely divided scales.

The wood work is of dull polished mahogany and the metal parts are simple in design but accurately made. No levelling screws are provided, as the instruments will work on any somewhat unlevel table without adjustment, wood wedges are however sent with them so that they can be levelled by the Quincke method if desired.

The designs of most of these instruments are registered.

No. 580. School Galvanometer.

The **School Galvanometer** is so constructed that it can be entirely dismounted, that it can be entirely taken apart and, whilst explaining the action of every part, be put together in front of the scholars somewhat after the following manner:

One of the magnets*) is pushed into the lower carriage on the suspension: Compass or Magnetometer.

One coil is placed parallel to the magnet: simple Galvanometer. The other coil is similarly adjusted and the windings coupled in series or parallel, finally the two coils are moved symetrically to the magnet: Varying degrees of sensitiveness.

The copper sheath is inserted inside the coils: Vibrations damped.

The windings of the two coils are connected in opposite directions: Differential Galvanometer.

The other magnet is inserted in its carriage: Astatic Galvanometer.

The pointer is adjustable as regards the magnet it can therefore always be arranged to point towards the scholars after the coil windings have been placed parallel to the pole axis of the magnet or in the magnetic meridian.

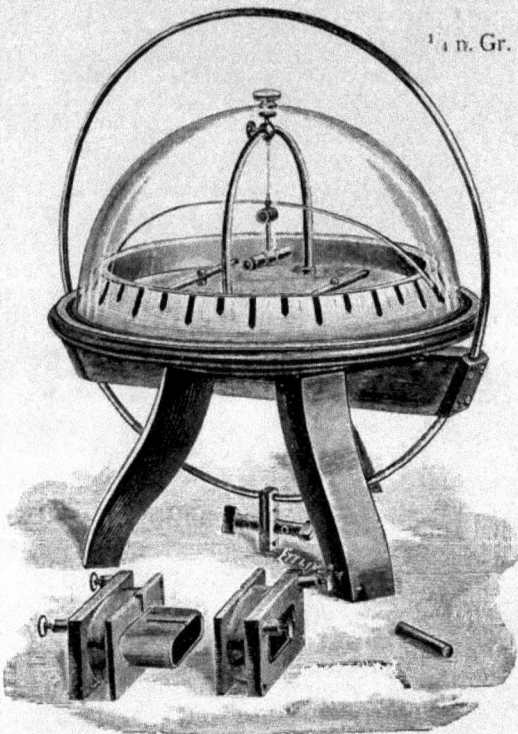

¹/₄ n. Gr.

No. **580.** **School Galvanometer** with scale graduated in divisions of 5 degrees for the scholars and a second scale graduated in single degrees for the teacher's use; sensitiveness with one magnet, therefore not astatic, with the coil windings in series 1 degree deflection = 0,00004 ampère. Price M. **75.—**

No. **580a.** **Tangent galvanometer ring** in copper mounted on wood tripod stand, the galvanometer base fitting centrally on the top of stand. Price M. **30.—**

When set up as Tangent galvanometer the instrument is suitable for measuring currents up to 15 ampères (10 ampères = 60° deflection), to enable the teacher to make more accurate measurements the large pointer can be replaced by a smaller one to read on the more finely divided scale.

*) The magnets are made of thin steel tube, both to reduce the moment of inertia and to render them as clearly visible as possible.

No. 581. Measuring wire for resistance bridge.

A measuring wire of constantan, 1 metre long is stretched between two double terminals over a scale divided into millimetres. The side of the base faceing the scholars is divided by wide black lines on a white ground into spaces of 5 centimetres with the central 4 decimetres into single centimetres.*)

$\frac{1}{6}$ n Gr.

No. 581. Measuring wire for resistance bridge; Range, using the galvanometer described overleaf and the resistances illustrated below 0.01 to 100 ohms. Price M. 30.—

One terminal at both the right and left hand ends of the measuring wire is connected to the poles of a battery (1—3 cells); one terminal of the comparative resistance is connected to an end of the resistance to be tested and the two remaining free ends are connected to the other terminals of the measuring wire, the galvanometer being connected between the junction of the unknown and the comparative resistances and the sliding contact.

No. 582. Comparative resistances for school work.

These comparative resistances are made with non inductive, free stretched wires. The smallest resistance of 0.1 ohm is made of 3 rows of copper wire, the 1 ohm resistance is in one row of German silver and the 10 ohm resistance is in ten rows of the same material, all the wires are of the same diameter so that the varying conductivity and the ratio between length and resistance can be clearly demonstrated.

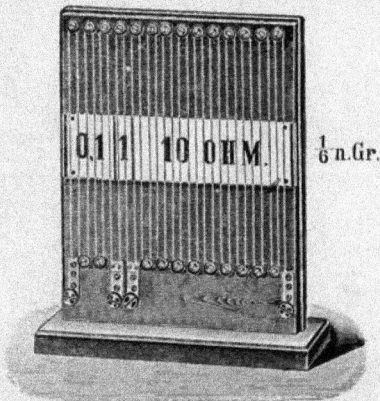

$\frac{1}{6}$ n. Gr.

No. 582. Three comparative resistances of 0.1, 1 and 10 ohms.**)
Price M. 25.—

*) Not shown in illustration.
**) The resistance of 10 ohms is now divided in two parts and fitted with a branch terminal for 2×5 ohms.

No. 583. School Voltmeter.

The principle of the Kohlrausch spring galvanometer must certainly be considered the simplest design on which to base the construction of a direct reading galvanometer. An iron core is sucked into a solenoid; the tension it exerts on a spiral spring supporting same acts as counterforce; the distance it is drawn into the solenoid is a measure of the pressure at the terminals, relatively of the current passing through the solenoid. To magnify this small movement for demonstration purposes, both the iron core and the spring are mounted on a lever which is connected to a moveable arbor and pointer.

No. 583. **School Voltmeter,** with a range of 0.5 to 3 volts, suitable therefore to measure the E. M. F. of single galvanic cells elements, with extra resistance to increase the range to 30 volts.   Price M. 40.—

No. 584. School Ammeter.

The construction of this ammeter is similar to that of the voltmeter, the principal difference being in the winding of the solenoid and the calibration. In both instruments the current is brought to the terminal at the left hand side and to the terminal in the middle of the base. Whilst the sensitiveness of the voltmeter is varied, or the range extended, by removing the coupling piece between the middle and the right hand terminal, in the ammeter the shunt is connected by inserting the coupling piece between them. If the approximate pressure or intensity is not previously known the measurements should always be commenced with the least sensitive arrangement of the instrument, in other words with the series- or shunt-resistance in circuit.

No. 584. **School Ammeter** with a range of 0.2 to 2 ampères, which by means of a shunt of one fourth of the resistance of the solenoid can be extended to 10 ampères.   Price M. 40.—

Millboard covers are supplied with both these instruments.

No. 585. Induction Apparatus.

The coils in this instrument are very carefully insulated; the secondary coil is moveable. A condenser which can be switched in and out of circuit is fitted in the base. The instrument is applicable to a great many purposes. Using current supplied from a few cells it is well adapted to generate alternating currents, such as are required when using a telephone to measure the resistances of electrolytes, it vividly incandesces

ca.
$^1/_6$ n. Gr.

Geissler tubes and answers the purpose of a transformer to demonstrate the conversion of a low into a high pressure current using small incandescent lamps; finally it can be used in combination with the school galvanometer for the Faraday experiments in induction, for which purpose a bar magnet is supplied to replace the iron core.

No. **585. Induction coil with condenser.** Ratio of the winding on the two coils about 1 : 10. Price M. **80.—**

No. 586. Telephone for demonstration purposes.

ca.
$^1/_6$ n. Gr.

This telephone comprises all the essential parts, magnet, induction coil and diaphragm all visibly arranged and easily taken apart. It can be used as an alternating current instrument for measuring the resistances of liquids, for which purpose a galvanometer is not per se a suitable instrument.

The method of connecting the unknown and the comparative resistances to the bridge remain the same, but the secondary coil terminals of the induction apparatus are connected to the bridge terminals instead of the battery.

No. **586. Telephone for demonstration purposes.** Price M. **15.—**

INDEX.